Lubrication of Electrical and Mechanical Components in Electric Power Equipment

Lubrication of Electrical and Mechanical Components in Electric Power Equipment

Bella H. Chudnovsky

CRC Press
Taylor & Francis Group
Boca Raton London New York

CRC Press is an imprint of the
Taylor & Francis Group, an **informa** business

CRC Press
Taylor & Francis Group
6000 Broken Sound Parkway NW, Suite 300
Boca Raton, FL 33487-2742

© 2019 by Taylor & Francis Group, LLC
CRC Press is an imprint of Taylor & Francis Group, an Informa business

No claim to original U.S. Government works

Printed on acid-free paper

International Standard Book Number-13 978-0-367-19745-2 (Hardback)

Library of Congress Cataloging-in-Publication Data

Names: Chudnovsky, Bella H., author.
Title: Lubrication of electrical and mechanical components in electric power
equipment / Bella H. Chudnovsky.
Description: First edition. | Boca Raton, FL : CRC Press/Taylor & Francis
Group, [2019] | Includes bibliographical references and index.
Identifiers: LCCN 2019010366 | ISBN 9780367197452 (hardback : acid-free paper)
| ISBN 9780429243035 (ebook)
Subjects: LCSH: Electric machinery--Maintenance and repair. | Electric
power-plants--Equipment and supplies--Maintenance and repair. |
Lubrication and lubricants.
Classification: LCC TK2189 .C455 2019 | DDC 621.31/042--dc23
LC record available at https://lccn.loc.gov/2019010366

Visit the Taylor & Francis Web site at
http://www.taylorandfrancis.com

and the CRC Press Web site at
http://www.crcpress.com

Contents

Preface

Electrical generation, transmission, and distribution machines are made of a complex electrical and mechanical components composed of various materials. These components are exposed to multiple mechanical and electrical stresses, as well as to environmental stresses (atmospheric corrosive gases, contaminants, and high and low temperatures). Multiple internal and external impacts result in progressive damage and significant deterioration of the electrical systems, which sooner or later cause disruption or even disability of the electrical apparatus to function properly and safely.

In electromechanical devices, the lubricants play many important roles. Lubricants separate the working surfaces, preventing metal-to-metal contact, and reduce friction and wear. Lubricants protect against corrosion, exclude contaminants, and dissipate heat. Reducing friction and corrosion through proper lubrication is equally important for electromechanical equipment with moving parts and for nonrotating apparatus. Proper lubrication materials and lubrication techniques not only provide safe and reliable equipment performance but also extend the useful life of power equipment.

This book is dedicated to research, engineering, and technical data in the field of electrical equipment lubrication. It presents analysis of many important applications of lubricants in the power industry of both electrical and mechanical parts. The book covers the major requirements of lubrication of mechanical components in two fields of electrical equipment—traditional and renewable energy generation by wind turbines.

The book reviews the use of lubricants for surfaces of electrical and mechanical parts to protect from mechanical wear and/or friction, and from degradation due to fretting. It analyzes the ways a lubricant serves as a barrier between metallic surfaces and atmospheric pollutants to protect the parts from corrosion. It underlines a danger of using of inappropriate lubricants, which may induce a significant risk by producing degradation, collecting particulate matter, and developing a high electrical resistance of the contact surfaces.

This book analyzes the effect of chemical composition and consistency (fluids, greases, solid lubricants) on the durability of lubricants in regards to various types of contacts and mechanical parts material, design, and load. The book is focused on the importance of carefully choosing lubricants to maintain stable contact resistance, preserve the physical integrity of the contact surface, and extend the useful life of mechanical parts, such as bearings. It presents a comprehensive list of lubricants manufactured worldwide, which are recommended by lubricants' manufacturers for and widely used in electrical industry including the renewable energy sector.

The book may serve as a reference material and a working manual for electrical, mechanical, safety, and maintenance engineers, as well as for training personnel. This book provides extensive information to aid in the proper choice of lubricant products and maintenance of electrical equipment.

About the Author

Bella H. Chudnovsky earned her PhD in applied physics at Rostov State University (RSU) in Russia. For 25 years, she worked as a successful scientist for the Institute of Physics at RSU, and since 1992 at the University of Cincinnati. Then she worked as an R&D engineer for Schneider Electric-Square D Company for 12 years, where her principal areas of activity were aimed at resolving multiple aging problems and developing means of mitigating deteriorating processes in power distribution equipment. In this field, she has published 40 papers in national and international technical journals and conference proceedings on topics that are summed up in her books. Her latest book, *Electrical Power Transmission, Distribution and Renewable Generation Power Equipment: Aging and Life Extension Techniques*, Second edition, published in 2017, is a review of the issues of aging and means of life extension techniques for energy power equipment.

Lubrication of Electrical Components

1

1.1 ELECTRICAL CONTACTS IN CIRCUIT BREAKING AND SWITCHING EQUIPMENT

Circuit breaking and switching equipment includes circuit breakers, switches, contactors, and relays.

Circuit Breakers. According to the American National Standards Institute (ANSI), a circuit breaker is "a mechanical switching device, capable of making, carrying and breaking currents under normal circuit conditions. Also capable of making and carrying for a specified time and breaking currents under specified abnormal circuit conditions, such as those of a short circuit". Slightly different is the National Electrical Code (NEC) definition of a circuit breaker: "A device designed to open and close a circuit by non-automatic means, and to open the circuit automatically on a predetermined overcurrent without damage to itself when properly applied within its rating." Since all circuit breakers (CB) belong to the same family of devices, they are much the same in general terms; they all are protective devices, used to interrupt current when an overcurrent fault occurs. There are a number of significant differences between the many types of electrical circuit breakers installed in various types of facilities today.

Relays, Switches, Contactors. A relay is a switching device; it can be controlled remotely and used to make a small signal control a larger signal (amplification of sorts). It is used to control currents in normal operation. A switch is used for switching an electrical power circuit, similar to a relay except with higher current ratings and a few other differences. When a relay is used to switch a large amount of electrical power through its contacts, it is designated by a special name: contactor. Contactors typically have multiple contacts, and those contacts are usually (but not always) normally open, so that power to the load is shut off when the coil is de-energized. Therefore, a contractor is a switch that is controlled by a circuit that has a much lower power level than the switched circuit. Contactors and relays are electromagnetic switches and operate under similar principles.

Electrical contacts in very-high-power circuits with currents up to mega-amperes do not have much in common with electronic circuits with currents as small as microamperes [1]. Power dissipation ranges from dozens of watts to a tenth of a watt [2]. The force holding the contact surfaces together may vary from low values ($P < 0.2$ N) to high (several tens of N) [3]. Many factors can lead to the degradation of contact surfaces. Sliding and rolling contacts constantly move against each other, and the surfaces degrade due to friction and wear debris. Static or wiping contacts experience vibration while the current is passing through, which destroys the plated coating (fretting). Nonconductive deposits on the contact surfaces formed due to various types of corrosion (atmospheric and galvanic) lead to a rise in contact resistance.

1.1.1 Factors Affecting Choice of Contact Lubricants

Lubrication provides better performance and protection of electrical contacts. In most applications lubrication is advantageous to electrical contacts because of lower insertion forces, less wear, and with special additives, also less corrosion. On the other hand, lubrication has disadvantages as well. Lubricated surfaces may retain more dust, the wiping motion may get less effective, a lubricant may enable sliding micromotion where this would not take place without lubrication), lubricant can also polymerize and form insulators, it can form a varnish, and it can creep to places where it is not wanted [4].

The positive effect of lubrication can be achieved only if the product is carefully chosen based on conformity between lubricant properties and the conditions of application, such as the contact design, material of the contacts, load, heat dissipation, and environment. If the lubricant cannot withstand the service conditions and degrades itself, inducing an additional cause of contact deterioration, the long-term effect of lubrication could be negative.

Contact Design. Lubricant function depends on application. In *separable* connectors, it reduces friction during installation, minimizes mechanical wear during connector service, and slows down the destructive effect of fretting corrosion. In most applications, lubrication is advantageous to electrical contacts because it provides lower insertion forces and less wear. Solid and liquid lubricants are often used to enhance the performance of electronic and electrical connectors. In the connector industry, one of the solid lubricants often used in power contacts is lamellar graphite, particularly with electrical brushes in machinery and electric trains, to take advantage of the unique lubrication and electrical conductivity properties of that material. Liquid lubricants reduce friction and minimize mechanical wear; they also must mitigate fretting corrosion and protect against atmospheric corrosion in contact areas of connectors [5]. The results of series of cycling tests showed that all lubricants had a beneficial effect on the performance of bolted joints as manifested by stable operation [6]. In contactor applications, the insertion and withdrawal force can be reduced to as low as 20% of the non-lubricated surface [7].

Contact Materials. The surfaces of noble metals are fundamentally different from those for non-noble metals. Noble metals have a clean and smooth surface and a pore-free layer if corrosion is a concern. At longer term the major concern with noble-metal contacts is to avoid pore corrosion and keep the surface free from contamination. A lubricant may well protect gold plating from corrosion Non-noble contacts have an oxide film on the surface and need much more force, 5–10 N, to disrupt the oxide film and make good contact [4].

Lubricants with special additives protect the contacts from corrosion. An important requirement for effective lubricant action in connectors is chemical inertness toward the metal surfaces to be lubricated. For example, lubricants used on copper- or silver-plated surfaces should not contain sulfur-containing compounds because these may react with the metals to generate resistive metal sulfide films and increase contact resistance. A lubricated contact surface reduces the friction force and the wear of gold plating during sliding or micromotion.

A summary of the last 20 years of lubricants testing for their application to various types of electrical contacts made of different types of materials is presented in [8]. The evaluation of the lubricants was performed on the basis of their effect on the performance and stability of the contacts, protection from atmospheric and fretting corrosion, reduction of friction and wear, and ability to withstand degradation under the contact working conditions.

Many lubricants have been tested on a bolted joint under various current cycling conditions and fretting to determine the spreading tendency, stability to thermal degradation, stability to UV radiation, and field conditions, as well their ability to protect the contact against fretting [9]. Detailed information on the requirements, methods of application, and examples of widely used fluids, greases, and solid lubricants may be found in [10].

> *Environment.* The lubricant in a permanent electrical connection blocks the access of corrosive gases or particulate material to the interfaces of the conductors [11]. An important factor in choosing the right lubrication product is the environment in which a particular electronic or electrical apparatus is to function. Various environmental factors and their combinations are aggressive toward traditional metals used in electrical contacts (copper, aluminum, and various alloys) and their coatings (silver, tin, nickel, gold, etc.), but they may also be harmful to lubrication materials.

1.1.2 Roles of Electrical Contacts Lubrication

In an electromechanical device the lubricants play many important roles. Lubricants separate the working surfaces, preventing metal-to-metal contact, and reduce friction and wear. Lubricants protect against corrosion, exclude contaminants, and dissipate heat. Reducing friction and corrosion through proper lubrication is equally important for electromechanical equipment with moving parts and for nonrotating apparatus. A power-switching device is an example of such apparatus. Its operation depends upon friction-free movement of many different mechanical components, usually through only a short distance or slight angle. These devices are often exposed to moisture and environmental contamination, as well as to long periods of idleness during which the lubricant deteriorates and materials corrode. Circuit breakers, which must operate quickly and reliably during unpredictable emergencies, are especially vulnerable.

The general principle of lubrication is to use a substance that protects an electrical connection from degradation and an increase in electrical resistance. Care must be taken to select the proper lubricant to achieve the desired results. In choosing grease, the three primary elements that go into the making of grease should be considered: the liquid lubricant (petroleum or synthetic fluid), the thickening agent, and additives. The chosen grease must be heavy enough to provide the right film strength yet be light enough to flow well in cold temperatures.

A properly chosen lubricant may slow down *degradation* without interfering with the electrical resistance of the contact. Application of lubricants is an effective way to protect an electrical connection from the factors that cause an increase of electrical resistance. In electrical and electronic devices, there are benefits from applying lubricants to the electrical contacts. The type and function of the contact define the lubricant required. By definition, an electrical contact is a combination of two surfaces held together by force with the passage of current, voltage, or signals [12].

A lubricant can also serve as a *coolant, prevent corrosion, and block the entry of contaminants*. Satisfactory grease will also resist separation, be reasonably stable, and not significantly change its nature within the temperature range for which it was designed [13–16].

Lubricants improve the performance of electrical apparatus in many ways, such as reducing mating forces, extending plating durability, and enhancing corrosion protection (fretting, galvanic, and harsh environments). Good grease will flow into bearings when applied under pressure and remain in contact with moving surfaces. It will not leak out from gravitational or centrifugal action, nor will it stiffen in cold temperature where it unduly resists motion [17,18].

Lubricants *prevent environmental and galvanic corrosion* on electrical contacts. Airborne contaminants attack metals resulting in corrosion. The types of corrosion materials depend on the type of contact metal and the environmental conditions, particularly gases that are in contact with the contacts. Contact surfaces and switches made of dissimilar metals are especially susceptible to moisture, oxygen, and aggressive gases. Corrosion causes oxides and other nonconductive compounds to gradually build up in pores until they reach the surface, where they impede current flow. Even noble-metal plating is at risk if it is worn or porous. The purpose of lubricant is to prevent dust, water, dirt, and other contaminants from entering the part being lubricated. A thin film of lubricant seals pores in plating and guards against scratches and substrate oxidation.

Lubricants *reduce wear and heat* between contacting surfaces in relative motion. Wear and heat cannot be completely eliminated, but they can be reduced to negligible or acceptable levels, especially on sliding electrical contacts where the lubricants separate the surfaces during operation and keep debris out of the contact area. Proper lubricants provide a balance between preventing wear and maintaining electrical continuity. A primary purpose of lubrication is to reduce friction between moving metal surfaces. To keep metal surfaces separated, lubricating grease should wet the metal and resist being displaced by the pressures it encounters. Because heat and wear are associated with friction, both effects can be minimized by reducing the coefficient of friction between the contacting surfaces. Lubricants reduce friction between mechanical components, which results in reducing the amount of force needed to activate a swift motion. Lubricants usually ensure a coefficient of friction of 0.1 or less, which means it takes little force to operate a device with a high preload. This can be important in switches and CBs where high normal forces ensure low contact resistance (CR) and a stable signal or power path. Lubrication is mechanically important because it gives the end user smooth, uniform operation.

Another benefit of using contact lubricant is that electric arcing during the switching operation is impeded by the dielectric strength of the lubricant, which is higher than that of air. In lubricated contacts, electric arcing occurs only when the distance between the contacts is already very short. When opening the contact, the electric arcs will cease much sooner. Shorter electric arcing times mean a general *reduction of contact erosion* and hence longer contact life.

1.1.2.1 *Protection from mechanical wear, friction, and fretting corrosion*

Lubrication is essential to achieve low and repeatable coefficients of *friction*, whether using inlay or electroplated material. The effectiveness of lubricants in stabilizing contact resistance, in wear reduction, and in retarding the rate of oxide formation by shielding the surface from air appears to be not strongly dependent on their composition and viscosity, and both fluids and greases are of value [19].

The microscopic wear particles oxidize quickly, turning into insulators. Build-up of oxide particles also accelerates wear. In general, hydrocarbon lubricants work best at wear prevention because their molecular structure is more rigid than other base oils. Some lubricants (polyphenyl ethers) appeared to be more effective than others in preventing adhesive wear, particularly when they are thick [19,20]. Mineral oil and POE are not very effective in reducing the coefficient of friction, probably due to their lower viscosity [20].

In a series of tests on tin- and silver-plated stationary parts, it was shown that lubricants generally reduced the severity of *wear* and tendency for *fretting corrosion* (thereby reducing the contact resistance), and gave no increase of CR or any other negative effect [21]. It was demonstrated in modeling experiments that lubrication might be effective in restoring a degraded system [22]. However, it was confirmed that lubrication delays but does not totally prevent fretting corrosion. After prolonged exposure to fretting, the beneficial effect of lubrication may be lost [23,24]. Further conduction is impeded by the formation of a layer composed of nonconducting wear debris and lubricant [25].

Fretting tests showed that at low loads none of the lubricants could protect the contact zone against fretting, but increasing the contact loads significantly improved the protective ability of the lubricant. This improvement was observed in copper-to-copper and copper-to-tin plated copper contacts, but not in the case of aluminum fretting against aluminum, since none of the lubricants tested had the ability to suppress the effect of fretting.

The results of the multiple fretting tests showed that many contact compounds have beneficial effects on the CR behavior. Application of a lubricant helps to decrease CR fluctuations and to lower the CR. The efficacy of a compound in reducing the fretting damage depends on the ability of a particular compound to flow back over freshly exposed metal pushed aside by the contact slip [26]. This ability depends on the lubricant's initial consistency and the change in consistency.

1.1.2.2 Shield from environmental corrosion

Corrosive factors in the environment such as corrosive gases and vapors are chemically aggressive toward contact materials and plating and produce a nonconductive deposit on the contacts. Other environmental factors, such as temperature extremes and humidity, accelerate corrosion. An additional contaminating factor is dust.

The benefits of applying lubricant for corrosion protection are not as well understood as its use in the reduction of friction and wear. For many years, the benefits of using lubricant for corrosion inhibition were overlooked because prospective users were skeptical that a nonconductive substance could be applied onto contacts without interfering with the conduction [27].

It is important to consider all the factors that may result in the negative effect of lubrication, for example if the chosen product deteriorates while exposed to a high-temperature environment, migrates or leaks out, or collects dust [28]. The choice of lubricant for corrosion protection should be based on thorough qualification of the product for survival in long-term use and the ability of such products to provide friction and wear reduction. Very few lubricants have been identified that provide exceptional long-term corrosion protection and yet do not produce any adverse effect on connector surfaces.

Field testing showed that lubricated surfaces collect and retain dust. Sliding electric contact experiments show that, in dusty environments, liquid lubricants appear to perform better than wax lubricants. Experimental and theoretical analyses indicated that the high permittivity of the lubricants plays an important role in attracting dust [29].

The mixture of dust particles and wax could be very harmful to the contact, especially under low normal force. It was found that liquid lubricants with lower permittivity provide much better contact behavior in dusty environments, either for sliding electric contacts or static contact applications [82]. There was little indication of any adverse electrical effects for the best lubricants. It appears that while dust may be retained by the lubricant, it is also effectively dispersed away from the mating interface during wipe or micromotion events. The tests found that the beneficial effects of corrosion inhibition by the lubricant overshadow any negative effects due to dust [29,30].

An example of a lubricant that proved useful for corrosion protection for connectors made or plated with precious metals is six-ring PPE or its mixture with microcrystalline wax, although the wax content should be controlled [24,25,31]. Multiple products, such as synthetic soap greases and other commercial greases, have been tested and proven useful for bolted contact application.

1.1.2.3 Inhibiting galvanic corrosion

Galvanic or bimetallic corrosion is another phenomenon that is harmful to the contact surfaces. Whenever dissimilar metals are in the presence of an electrolyte, a difference in electric potential develops. When these metals are in contact, an electrical current flows, as in the case of any short-circuited electric cell. This electrolytic action causes an attack on the anodic metal, leaving the cathodic metal unharmed. The extent of the attack depends on the relative position of two metals in contact in the electrolytic potential series.

Galvanic corrosion is common with aluminum-to-copper connections, since copper and aluminum are quite far apart in the series, copper being cathodic and aluminum being anodic. Hence, when aluminum and copper are in contact in an electrolyte, the aluminum may be expected to be severely attacked.

Some products, such as petrolatum-type compounds containing zinc dust, effectively protect contacts made of dissimilar metals from galvanic corrosion [32]. However, some lubricants may induce galvanic corrosion; therefore the products for this application should be chosen very carefully. For example, graphite, which has a noble potential, may lead to severe galvanic corrosion of copper alloys in a saline environment [33].

Multiple studies showed that lubrication reduces the ingress of corrosive gases on the contact metals or through pores (which is particularly important for thin gold plating) to react with the substrate metal. Application of lubricant reduces the possibility of electric contact failure from insulated particulates such as dust or wear fragments of tarnished films, which may deposit on the contact area.

The most important and general roles of the lubrication of electrical contacts are protection from environmental, galvanic, and fretting corrosion as well as from wear and friction. All known types of lubricants (among them mineral and synthetic oils and greases, solids, and dispersions) [34] have been tested and applied to electrical contacts. Compared with other lubricants, polymeric ethers such as PPE and Perfluorinated Polyether (PFPE) have better characteristics: stability at high temperatures and a very low vapor pressure. Above all, PPEs demonstrate an ability to remain at the contact point without migrating.

However, application conditions of connectors (such as electric current, normal force, the number of contact pairs within one housing, sliding, micromotion, etc.) and environmental conditions (such as humidity, temperature, corrosive gases, dust, etc.) are very complicated. That is why it is almost impossible to choose just one kind of lubricant to fulfill all the requirements and to match all the above conditions. Unfortunately, there is no such thing as a "universal lubricant."

1.1.3 Lubrication Failure Modes in Contact Applications

The term "lubrication failure" includes different issues, such as inadequate amount of lubricant applied or no lubricant at all, improper product chosen for lubrication, contamination and aging of the lubricant because of ambient conditions, and so on. It is found that as many as 60%–80% of all bearing failures are lubrication related, whether it be poor lubricant selection, poor application, lubricant contamination, or lubricant degradation. Many components are failing early because lubrication best practices have not been established.

When failure happens, a thorough examination of the lubricant's properties is required in many cases, as well as analysis of the amount of lubricant applied, where it was applied, and the operating conditions of the mechanical or electrical parts. All these factors contribute to the damage caused by "lubrication failure," which led to excessive wear and corrosion of parts and the failure of the electromechanical unit to perform properly [35].

Lubrication failure may affect the functionality and performance of various parts of electromechanical apparatus. The role of the proper choice of the lubrication products, terms, and procedures is usually discussed in relation to operational and environmental parameters. Lubrication failure mode describes the specific causes of failure associated with a component or functionality of a process. The failure may be caused by lubrication procedures, terms, or products. For example, the mode for circuit breaker (component) failure to open (functionality of a process) is lack of lubrication.

A number of different factors may affect lubricant performance. It is important to choose the correct product for each particular application, which means that many factors should be considered, such as operational and thermal temperature ranges, mechanical load (pressure) on the areas where the lubricant is applied, presence of moisture, possible contamination (particulate matter, dust, corrosive gases), and so on. Accordingly, the lubricant properties (working temperature range, moisture content, and presence of corrosive chemical components) should comply with the application parameters and materials to avoid adversely affecting the units and environment.

Knowledge of the composition of the lubricant is particularly important for its application to electrical contacts, since some additives may affect electrical resistance of conductors. It is also important to apply the right amount of lubricant, because the lack of it or an excessive amount may increase the friction and wear of the parts or lead to contamination of other parts of the unit with lubricants.

1.1.3.1 Thermal failures

Major factors in the durability of many lubricants are thermal limitations, both high and low.

Grease has a maximum temperature at which it can safely be used. This critical temperature is called the *drop point* at which the gel structure breaks down and the whole grease becomes liquid. When grease is heated above its drop point and then allowed to cool, it usually fails to fully regain its grease-like consistency and its subsequent performance will be unsatisfactory. Accordingly, if the temperature of CB lubricated components in operation is higher than the upper limit of the lubricant working temperature range, it could leak out at high temperatures, which will affect the mechanical performance of the parts or expose them to corrosive environment. General lubricant terminology is given in Addendum 4, Lubrication Glossary.

An elevated operating temperature shortens the life of greases. In fact, if a switchgear operates at temperatures above 70°C (158°F), it cuts grease life by a factor of 1.5 for each 10°C rise [36]. High temperature promotes oxidation and increases oil evaporation rates and oil loss by creep, which accelerate grease drying and shorten grease life. Grease life can be estimated for operating temperatures above 70°C (158°F) with moderate loads and no contamination. This approach is applicable to fresh industrial greases of Grade 2 consistency with thickeners such as lithium, complex metal soaps, and polyureas.

Other ways to estimate lubricant lifecycles are based on operating conditions and require the application of factors to account for real-life conditions, such as solid contamination, moisture, air, catalytic effect of wear debris, temperature variations in a circulating system, and so on. As the temperature increases, the rate of grease degradation increases. The Arrhenius rule suggests doubling the lubricating oil degradation rate for each 10°C increase in temperature. Synthetic lubricants generally last longer at elevated temperatures than their mineral-oil counterparts.

The base oil of the grease determines the minimum working temperature for the grease. The base oil of the grease for low-temperature service must be made from oils having a low viscosity at that temperature. For example, if the operational temperature of the unit in operations is lower than the lower limit of the lubricant working temperature range, then the lubricant may stiffen at cold temperatures, which could happen in the CBs' outdoor installations [37].

1.1.3.2 Degradation of contact lubricants

The purpose of lubrication is to improve electrical-contact functionality and prolong service life. Therefore, it is important to ensure that none of the lubrication product properties would work against this purpose. Modern lubricant used in electrical industry for lubrication of electrical contacts is often a complex mixture of solid, liquid, and polymeric ingredients. It also includes multiple additives improving various lubricant properties, such as thermal stability and corrosion resistance. Using such complex dielectric materials on a conductive surface may cause a serious problem affecting performance of the contact.

Lubrication products should not deteriorate or contaminate the contact surfaces for a predetermined period of time within the range of application conditions. A lubricant should remain stable in sufficient quantities at the contact interface to perform its intended function throughout life.

Lubricants applied to the contact surfaces are exposed to four major temperature-sensitive aging mechanisms: evaporation, surface migration, polymerization, and degradation [24,28,38]. With time and temperature, lubricants may degrade by oxidation or polymerization, forming insulating films or gums [39].

Due to evaporation and mechanical loss, a thin coating of a contact lubricant can disappear from surfaces. Mechanical loss is a removal by sliding which results in wear debris carrying the lubricant away. Evaporative loss will occur if the lubricant has a significant vapor pressure and is important at elevated temperatures, particularly with the fluid lubricants [24].

When a lubricant disappears from the contact interface by evaporating or spreading, the contact is exposed to failure by wear or by corrosion. Several material properties, such as viscosity, surface tension, vapor pressure, and thermal stability, are important in selecting the proper lubricant for contact application. Thermal limitations, both high and low, represent major factors in the durability of many lubricants. At low temperatures, many lubricants appear to solidify, developing high-shear-strength films, and leading to high

CR [40]. Some lubricants are susceptible to cracking due to long-term exposure to high temperature [21].

Lubricants may degrade due to chemical reaction with the atmosphere and may also polymerize when heated in the presence of a copper-based alloy [41]. Copper is known to degrade polymers through catalytic-enhanced oxidation, so grease would undergo changes in contact with copper at higher-application temperatures, which in turn induces copper corrosion. Applying grease with a copper corrosion inhibitor to copper contacts at high temperature may significantly decrease lubricant-induced copper corrosion [42].

Because of their composition, some lubricants, such as greases with silicate thickener, have undesirable properties when applied to electrical contacts [43,44]. Silicate as a high-temperature grease thickener may pose problems if entrapped under the contact. Near an arc, splattering effects can occur, coating the surface with a nonconductive layer. Particles may also be fused together by the arc. Even with lower voltages, arcing will occur if the inductance in the circuit is significant. Silicon compounds, both inorganic and organic, are detrimental contaminants for arcing electrical contacts [44]. Any refractory material in the area of arcing can present similar problems.

Another problem is a possibility of the liquid component of the lubricant to evaporate over time in service, which results in drier lubricant and the formation of a solid-like film. Such film is harder to displace under low normal forces. The risk is that a loss of weight from the lubricant could create an impenetrable film on the contact and interfere with the electrical connection across a contact interface. This problem was studied in Ref. [45] by conducting and analyzing the results of the tests using coupons plated with 15 micro-inches (0.38 μm) of gold plating and a proprietary TE Connectivity (TE) contact lubricant.

1.1.3.3 Effect of surface reactivity of lubricants on contact resistance

Stearic acid is an effective lubricant for aluminum sliding electrical contacts to minimize friction and mechanical wear and protect electrical interfaces against corrosion. Its lubrication properties are excellent because the long-chained acid molecules are chemically attached to aluminum oxide film naturally formed on the contact surface. However, the study presented in Ref. [6] shows the importance of differentiating between the lubrication and the "electrical-contact protection" properties of lubricants.

In an electrical contact, mechanical disruption of this native oxide film during sliding exposes the underlying aluminum metal to the stearic acid. The following chemical reaction of the acid with the exposed metal leads to chemical breakdown of the lubricant and immediate surface oxidation of the aluminum. This reaction causes a large increase in contact resistance. These observations provide clear evidence that stearic acid is unacceptable as a lubricant for aluminum electrical contacts, although it is effective as a conventional lubricant for aluminum.

1.1.3.4 Lubricant inducing contact corrosion

High performance switch grease (Li-stearate PAO grease with and without corrosion inhibitor) was tested for corrosivity to copper. The testing was performed according to by the standard ASTM 4048-91 Standard "Test Method for Detection of Copper Corrosion from Lubrication Grease." The test [46] involves totally submerging a test coupon in a large amount of the lubricant and heating the assembly at 100°C for 24 hours. After the test, the degree of corrosion noted by the color of a coupon showed that lubricant produced some tarnish.

However, in real application the copper contact is never submerged in grease: the grease would be applied to the contact surfaces in thin layers, much thinner than recommended in the ASTM method. When tested with the thin grease layers on copper surfaces, corrosion was more intense of that seen using the ASTM method [42].

This study revealed the fact that a copper contact surface might experience more accelerated corrosion when a thin layer of lubricant is deposited on the contact. A grease that passes the ASTM qualification test for copper electrical contacts produces a thicker contamination layer when aged with a thin grease layer (0.12 mm) over copper. The contamination layer significantly affects electrical performance determined in contact resistance measurements.

1.1.4 Lubrication of Contacts in Circuit Breakers

The lubricant choice for a specific application is determined by matching the machinery design and operating conditions with the desired lubricant characteristics [47]. Grease, not oil, is generally used for the lubrication of circuit breakers, equipment that does not run continuously and whose parts can stay inactive for long periods of time. Circuit breakers are not easily accessible for frequent lubrication thus using a lubrication grease is preferable. Grease should be chosen for lubrication of most parts of CBs, since they can be exposed to high temperatures, shock loads, or high speeds under heavy load. In the application to circuit breakers, choosing the right lubricant is a very important step. High-quality greases can lubricate the components for extended periods of time without replenishing.

If a lubricant containing not-conductive solid additives is applied to electrical-contact surfaces it will increase electrical resistance, leading to overheating failures. To avoid such problems, it is important to use both the type and quantity of lubricant specified by the manufacturer, not only at the time of installation but also during maintenance procedures.

The greatest risk to personnel presents when racking a circuit breaker, in or out. Unlubricated primary stabs increase the force required to connect them to a power source. On the other hand, as the lubricant on primary stabs gets gummy it will take more effort to rack in the breaker, increasing the potential of an arc flash incident.

Pivot points require lubrication. As a circuit breaker carries load, heat is produced by the current losses across the current path. The higher the load current, the greater the losses will be and the more heat will be produced. Heat increases the rate at which the lubricants dry out, even if the breakers are exercised regularly. Grease is forced out of pressure points over time, the exact points that need lubrication the most. Exercising the circuit breaker will move lubricants back into the pressure points. However, all lubricants will eventually dry out over time and need to be replaced.

Only a thin layer of lubricant on primary and secondary connection points is required, just enough to reduce the resistance of the mechanical connection. Excessive lubrication will eventually attract dirt and other deposits. On the other hand, lack of lubrication is the number one problem when servicing circuit breakers in the field. All circuit breakers have lubricant applied to them at the factory, but over the course of time this *lubricant dries out, gets gummy, and then flakes off,* leaving metal-to-metal wear in its place. Some of the most important points of lubrication are primary and auxiliary connections.

Unlike bolted pressure switches, circuit breaker contacts usually should not be lubricated. The heat from the current flow through the contacts breaks the lubricant down and causes contact resistance to increase. It may be acceptable to apply an extremely thin layer of lubricant to *protect from corrosion* in certain atmospheres [47].

Electrical contacts should not be lubricated with metal-filled lubricants unless tested and proved to be effective in the long term. Many can accelerate corrosion, create conductive paths, and eventually cause failure. The general rule is to avoid graphite, molybdenum disulfide (Moly), or PTFE (Teflon®) lubricants for electrical contacts, because they could cause a rise in resistance after multiple operations [48]. For most switches and the ones that operate infrequently, to just keep the contact clean and dry with no lubricant might be a viable option. Main and arcing contacts should never be lubricated. Proper lubrication practices for various types of low voltage and medium voltage power distribution equipment are presented in multiple papers [49–55] and OEM's data bulletins [56,57] available for customers.

1.2 LUBRICATION OF ELECTRICAL CONNECTORS

1.2.1 Electrical Connectors

An electromechanical device used to join electrical terminations and create an electrical circuit is called an electrical connector. Electrical connectors usually consist of plugs (male-ended) and jacks (female-ended). The connection may serve as a permanent electrical joint between two wires or devices. Connectors may join two lengths of flexible copper wire or cable, or connect a wire or cable to an electrical terminal.

Connectors for portable equipment could require a tool for assembly and removal. An adapter can be used to effectively bring together dissimilar connectors. There are hundreds of different types of electrical connectors, which are manufactured for power, signal, and control applications.

Application conditions of connectors are very complicated. There could be multiple conditions that vary widely, such as electric current, normal force, the number of contact pairs within one housing, sliding, micromotion, etc.

Environmental conditions such as humidity, temperature, corrosive gases, dust, and others may also vary significantly for different connectors' locations. Therefore it is almost impossible to choose just one kind of lubricant ("universal lubricant") to fulfill all the requirements and to match all the above conditions.

1.2.2 Failure Causes of Connectors

Most failures of portable and not-portable connectors are originated from the following causes.

Wear. The most common cause of failure in portable systems such as cellular telephones, computers, instruments, etc., is a defect at the metallic interface inside a connector. The metallic interface inside a connector is typically tin or tin-lead, but may be plated with a very thin layer of a noble metal such as gold. Whether a pin is tin or gold, the failure mechanism begins with wear due to friction at the metallic interface. In cellular telephones, wear is usually the result of micromotion, which may involve displacement between the pin and the socket of only a few microns. Micromotion in turn is caused by vibration from handling of the telephone and temperature change.

Environmental Corrosion. In connectors below the corrosion-resistant gold and tin layers lay less noble sublayers, which are exposed after multiple micromotions. These sublayers are quickly oxidized and are sensitive to contact with moisture and other airborne corrosive chemicals. The moisture and airborne contaminants (sulfuric acid, nitric acid, and many more) create electrolytic cells that result in corrosion of the pin causing rapid failure. Failure actually occurs when the corrosion products grow and eventually separate the metallic surfaces at the point of contact. Contact corrosion is a primary failure mechanism.

Contamination. Airborne contaminants are present everywhere, including dozens of various organic and inorganic contaminants in the form of suspended aerosols or fine particulate solids, which attack not only connectors but also plastic-packaged integrated circuits (ICs). In plastic ICs, contaminants and moisture may enter along the external leads, or they may settle on the package surface and then migrate as ions through the epoxy. In cellular telephones, oxygen, moisture, and contaminants penetrate to the connector along the surface of the pin but they also may migrate into the plastic housing around a connector and may eventually reach the metallic surfaces of the connector.

Fretting Corrosion. Another mechanism known as fretting corrosion leads to abrasion of contact surfaces. When a metallic surface is plated with gold, which is usually very thin—a fraction of a micron—plating may be removed; if the pin is plated with tin-lead, the surface is scored, which also may lead to the connector failure. Fretting and atmospheric corrosion are two mechanisms that lead to the failure of connectors in other than portable instruments.

Connectors for electromagnetic systems in aircraft, automobiles, and industrial plant may have to operate in the presence of high levels of humidity, pollution, and vibration and, if wrongly specified, may be particularly susceptible to fretting corrosion. This in turn may result in contact intermittencies that may have a deleterious effect on system performance: faults that appear and disappear more or less at random and are extremely difficult to diagnose either by bench testing or by in-service monitoring. Even the use of gold-plated contacts provides no absolute guarantee against fretting if the flashing is of insufficient thickness. With safety-critical electromagnetic components, electrical-contact degradation may have serious consequences. An improvement of contact reliability may be achieved by using proper electrical-contact lubricants [58].

1.2.3 Protecting Connectors from Failure with Lubricant

1.2.3.1 Purposes and forms of connector lubricants

Contact lubricants play three important roles in protecting electrical circuits [59]. First, they prevent damage to the contact surface from the nearly imperceptible movements that occur during operation. Second, lubricants seal the connection to stop moisture from entering and corroding the contacts. This is especially important for connections exposed to the elements. Finally, the lubricants reduce insertion forces to speed assembly and prevent workforce fatigue and injury.

The two biggest challenges the interconnect industry faces during product validation are high mating forces and low electrical performance. Most general lubricants have little or no positional stability and work briefly or not at all on electronic connectors. Contact lubricants should reduce initial mating wear and sliding friction and therefore preserve contact integrity and electrical performance.

Which lubricant to choose mostly depends on the type of contact finish with two basic types used in connectors: *noble*, represented by gold (Au), palladium (Pd), and some alloys of these metals; and *non-noble*, which are mostly tin finishes, whether plated or applied by controlled exposure to a liquid metal source such as Hot Air Leveled Tin (HALT). Consequently, the requirements on the lubricant and its formulation are different for noble and non-noble applications.

Gold finishes are typically used in connectors intended for more demanding applications, such as those requiring high performance/reliability, which in turn requires low and stable contact resistance. In such cases the lubricant must provide corrosion protection for resistance stability and reduced friction for reduced connector mating forces, with the coefficients of friction for gold and tin finishes are in the range of 0.3 and 0.7, respectively. The coefficients of friction of lubricated connectors can be in the range of 0.1, a significant reduction that is the source of the reduced mating force and wear rates [60].

The substance must lubricate the surfaces against micromotion and also against larger excursions. Connector lubricant must continue to provide effective lubrication over a fairly wide temperature range. Because the lubricant must provide connector reliability during the life span of the equipment, which may be several years, it must have a low vapor pressure (evaporate very slowly).

Other physical and chemical requirements for the material are that it should be stable when exposed to high temperature and oxygen. The lubricant must operate well on noble metals, including gold and palladium. It must also perform well on materials such as tin and less noble base metals. Finally, it must be chemically inert to the materials of the connector and the connector housings.

Lubricants are designed for various electrical applications. For example, in application in the automotive industry, the products must protect components from wear and corrosion. These products should provide protection against external conditions such as dust, moisture intrusion, humidity, high engine compartment temperatures, and long overnight exposure to very cold temperatures [up to at −40°C (−40°F)].

Lubricants should be chosen dependent on contact design (sliding, butt action, slip ring, etc.), contact forces, switching current requirements, and compatibility with plastics, elastomers, and metals. Low current switching with low contact forces may require a lubricant with different chemical and physical properties compared to switching high current with high contact forces.

Connector lubricants are available in three forms [61]:

1. *Oils*—most often applied as a field fix to correct a problem. They are occasionally used during production, where they are atomized and sprayed onto terminals as they are stamped.
2. *Greases*—injected into the female part of the connector. Greases can be applied both before and after a connection is made, and they can contain a variety of additives to solve specific problems.
3. *Dispersions*—consist of greases dissolved in a solvent to make them more liquid. They are easy to apply in the production environment by spraying or dipping. After a dispersion is applied, the solvent evaporates to leave a thin film of lubricant on the contact.

Connector lubricants are formulated from a variety of chemicals:

1. *Polyalphaolefin (PAO)*—the most common synthetic lubricant; typically known as an anti-fretting lubricant. It is available in a range of grease thickener systems and provides good protection at temperatures to 125°C or higher.
2. *Perfluoropolyether (PFPE)*—used in high-temperature applications; provides excellent protection at temperatures up to 250°C. It provides good insertion force reduction.
3. *Polyphenyl ether (PPE)*—typically used on gold contacts. It provides unique film strength capability that prevents galling of gold when the contact is made. PPE is the most expensive lubricant used.
4. *Specialty silicones*—used occasionally for contacts that require high temperature resistance above 250°C (482°F).

The following factors determine the choice of connector lubricants:

1. *Temperature range*—PAO used below 135°C, PFPE used above 135°C (275°F)
2. *Elastomer compatibility*—without actually testing a lubricant on a material it is virtually impossible to make accurate, all-inclusive compatibility recommendations. A single elastomer category, for example, Nitrile, can have as many as 100 possible formulations, each with different compatibility issues. Materials should be exposed to lubricants under various temperatures and loads during compatibility testing. The most accurate compatibility testing should be done at the expected operating extremes. For further assurance, one should test materials of nearby components in case of oil migration or outgassing and condensation.
3. *Insertion force reduction*—PFPE/PTFE provides the greatest reduction.
4. *Cost*—PAO provides the lowest cost where it can be used.

1.2.3.2 Lubrication reducing wear, friction, and fretting corrosion of connectors

The lubricants are basically insulators, but they still have a positive effect on the performance of electrical contacts. Several causes of failure of portable and not-portable connectors may be reduced with the help of lubrication. Electrical contacts are unlikely to meet the ever-increasing requirements of today's applications without a lubricant. Even electroplating or chemical coating of the contacts does not always have the desired or required effect. Also, the use of coatings can become prohibitively expensive if layers of a certain thickness are required.

When the electrical contact is closed, the lubricant forms a separating film due to the relative motion of the two contacts, despite the contact force pressing the contacts together. A properly selected lubricant lowers *insertion force* by decreasing the *coefficient of friction* between mating surfaces. It reduces *mechanical wear* by placing a film of oil between the mating surfaces. Due to the contact force, the lubricant is displaced from between the roughness peaks while the contact is stationary, so there is direct metal-to-metal contact between the surfaces. The result is a low electrical-contact resistance [62].

Lubricants often have to be used on electrical connections if the contact resistance and the actuating forces should remain constantly low for as long as possible under specific operating requirements.

Lubricants provide *wear reduction* when a high number of plug or contact cycles is required, such as in Smart Card connectors and plug-in connectors in automation technology.

Lubricants provide *friction reduction* when low plug and unplug forces are required, such as in back planes in telecommunications and multi-pin plug-in connectors for data lines. Lubricants help to reduce *fretting corrosion* where long service life and contact reliability are required as well as when contacts are operating under vibration and frequent temperature cycles (automobiles, automation technology). Coating the contacts with an anti-fretting lubricant reduces not only mechanical wear but also provides an oxygen barrier and helps keep oxide debris away from the contact area.

1.2.3.3 Lubrication minimizes environmental contamination of connectors

Contact surfaces can get coated by layers of foreign matter or can change chemically due to ambient and operating conditions of connections. The resistance is increased by the foreign layers that are frequently found on the contact surfaces, such as metal oxide layers or plastic deposits. It takes a sufficiently high contact force or heat generation due to power loss to penetrate these layers. The lubricant covers the open contact (during storage), thus preventing the formation of a detrimental layer of foreign matter such as an oxide on its surface.

The lubricants used to prevent connector corrosion displace moisture and provide an environmental barrier to protect contacts in a wide range of electronic equipment and mechanical components. Such lubricants, which meet the corrosion-related performance requirements of MIL-DTL 87177 B, are used in many demanding industries including aerospace, transportation, manufacturing and agricultural [63].

A very important field where connector failure is not acceptable is medical equipment, such as dialysis machines or in pulmonary monitors, which have to exhibit long-term and uncompromising reliability [64]. However, when a connector lubricant is used, it nearly always is applied only once—during electronic assembly of the system. Relatively few connectors in installed systems are readily accessible for easy lubrication, and few repair technicians carry connector lubricants with them.

When connector failure does occur, in most cases it necessitates costly equipment disassembly for the purpose of replacing the corroded component. Whichever type of connector lubricant is used in a particular medical system, the material must have several distinctive properties. Probably most important, it has to remain where it has been applied. The lubricant is useless if it migrates away from the connector, or, more specifically, from the two in-contact surfaces.

1.2.3.4 Connectors protected from abrasion and corrosion with polyphenyl ethers

It was found that connector's metallic interface can be protected from both abrasion and corrosion by a specialized class of lubricants called polyphenyl ethers (PPE), which have the characteristics that practically no other lubricants have: stability at high temperatures, very low vapor pressure, and above all, the ability to remain at the contact point without migrating. They remain stable at temperatures higher than those commonly encountered in cellular phones or other electronic systems. Chemically, they are relatively inert, and when used on connector pins would not react with nearby metal and plastic elements even if they could migrate to them. They also have a very low vapor pressure. A very thin film of PPE deposited onto a tin-lead connector pin will, at normal temperatures and pressures, evaporate only after 40 to 50 years.

PPEs provide a high degree of protection to connectors from atmospheric contaminants, though the mechanism of this protection is not yet fully understood [65]. It was determined that applied to connector pins in a very thin layer, PPE can extend the service life of connectors by a factor of 1,000 or more. Increasing the service life of connectors—the most frequent failure site in cellular telephones—greatly increases the reliability of the telephone.

A very interesting test was conducted by coating an already-corroded connector pin with PPE. The performance of the pin immediately improved because the droplets of the lubricant absorbed the corrosion products that were preventing contact between the pin and the socket. Studies have shown that PPEs provided the best performance on connectors; however, PPEs are expensive and have limited low-temperature operability, solidifying at 20°C (68°F).

Polyphenyl ethers meet all of the most critical requirements. An example of the superiority of performance offered by PPEs is that they remain stable and continue to act as effective lubricants at temperatures as high as 453°C (847°F), while petroleum-based and other synthetic lubricants decompose at around 200°C (392°F). PPEs are also extremely resistant to ionizing radiation, a property that has led to their widespread use in nuclear facilities and on earth-orbiting satellites.

Not all electronic connectors require the top-level lubricating performance of PPEs to achieve high reliability. But until recently, a wide performance gap existed between low-end connector lubricants and PPEs, and an assembler who needed superior performance had only PPEs to choose from. This gap has now been filled, however, through the development and introduction of advanced phenyl ether (APE) lubricants.

The single most important performance property of connector lubricants, as mentioned, is positional stability: the lubricant cannot protect the connector if it has migrated to a different location. APE lubricants have a surface tension that is lower than that of PPEs but significantly higher than the surface tension of ordinary connector lubricants. When applied to the connectors in, for example, an X-ray system or a ventilator, an APE lubricant ensures reliable equipment function by remaining dependably in place at the connector interface.

1.2.3.5 Application of lubricants to the connectors

Proper and controlled application of lubricant to the connectors is a very important step. When a lubricant is applied before assembly it may be partly removed during the assembly process, but after assembly the contacts are mostly less accessible. Housings and packaging materials may become contaminated with lubricant.

It is well known that the lubricant thickness is important for the electrical function. However it is hard to measure and hard to control the thickness initially and also after a few mating cycles have taken place. Lubricants must still be considered to be a part of the surface finish. Therefore it has to be specified on the product drawing and subjected to the procedures of testing and quality control, like any other finish.

Two different type of connector lubricants are manufactured: petroleum or vegetable-based oils that do not contain water, or "neat" lubricants and solvent dispersions containing a solvent or polymeric based binder. Neat lubricants tend to be used in larger connectors, where the larger volume of lubricant acts as an environmental seal. Solvent dispersions are recommended where only a thin film of lubricant is desired.

1.2.4 Lubricants for Automotive Electrical Connectors

1.2.4.1 Automotive connectors' challenges

The number of electrical systems in a typical passenger vehicle continues to grow, powering everything from headlights and DVD players to body impact sensors and global-positioning systems. With each new system comes additional connectors. Luxury cars, for example, now have more than 400 connectors with 3,000 individual terminals, which means there are 3,000 potential trouble spots.

Connectors have to be protected against water, changing temperatures, road grit, and vibration, all of which speed oxidation and fretting corrosion on connectors. Corrosion generates resistive oxides on the connectors, which cause intermittent faults and electrical failure. The challenge for connector manufacturers is to extend the operating life of their products as more are used on each car and car companies continue to extend their warranties.

Obviously, lubricants have an important role to play in connector performance, including corrosion prevention and cost reduction. This is especially true for low-voltage connectors (0.1 to 0.5 W), which now constitute about 75% of connectors in passenger cars. The right lubricant lowers the insertion force needed to assemble connectors, reduces mechanical wear by placing a film of oil between mating surfaces, and, if the lubricant contains the proper additives, minimizes corrosion. Such lubricants can be based on a variety of chemistries, but synthetic hydrocarbons and ethers currently dominate the market.

1.2.4.2 Lubrication of high-temperature terminals

Perfluoropolyether (PFPE)-based greases are among the most frequently chosen synthetic lubes in the automotive industry. Since PFPE greases do not oxidize at higher temperatures they are usually selected for high-temperature applications (up to 250°C). In the 1980s, lubricant engineers developed greases that combined the stability of PFPEs with polytetrafluoroethylene (PTFE) to make a lubricant with an exceptionally low coefficient of friction [66].

However, in automotive applications the high normal force and heat, combined with the action of putting connectors together and taking them apart several times, burnishes PTFE into the surface of the contact, insulating the contact asperities that actually carry current. This leads to intermittent and catastrophic connector failures. PFPE may be desirable as a base oil due to its high-temperature performance and resistance to oxidation, however another potential problem is that the inert chemistry and high specific gravity of PFPE make it difficult to thicken the grease. The solution was found where a soluble, slightly basic nitrogen compound—urea—replaces PTFE to avoid the burnishing that increases contact resistance. When urea is part of the mix, the grease separates at high temperatures due to density differences between PFPE and urea.

Other electrical contacts exposed to wide temperature ranges may be lubricated with ester grease with clay thickener, polyol ester grease with Li soap, or organic polymer thickener. However, these greases are not compatible with ABS, polycarbonate, polyester, PPO, or PVC plastics, or BuNa S (Bu for butadiene and Na for sodium, and S for styrene), butyl, or neoprene elastomers. The working temperature range for these greases is from −40°C to 149°C (−40°F to 300°F).

1.2.4.3 Lubricant compatibility with plastics

The choice of the lubricants for use in automotive electrical connectors depends on the specific application [67]. For both ferrous and nonferrous static electrical connectors, synthetic hydrocarbon with silica or Li soap thickener lubricants are recommended. These lubricants are compatible with most engineering plastics including polycarbonate and acrylonitrile butadiene styrene (ABS) (an opaque thermoplastic and amorphous polymer). Another choice for lubricants compatible with plastics is PFPE grease with inorganic thickener. This grease has low volatility and high oxidation resistance for extended durability in thin films and extremely wide operating temperature range from −70°C to 200°C (−94°F to 392°F).

For high current switching with low arc debris, glycol non-melting grease with inorganic and Li soap thickeners is recommended to protect Cu and Ag from corrosion, and provide good mechanical and chemical stability.

1.2.5 Use of Connector Lubricants in Avionics

1.2.5.1 Connectors' challenges in avionics

Avionics systems control the operation of flight-critical and flight-essential equipment, including navigation, communications, power distribution, flight and engine controls, displays, and wiring. These systems assume a major responsibility for the performance, safety, and success of commercial and general aviation. The reliability of these complex and often interrelated systems in any environment is critical for safe operation [68].

The wiring on aircraft can act as conduits for water condensed by changes in temperature and altitude during flight or on the ground. Once the condensation forms within the wire bundles it starts to travel to the lowest point in the harness—usually the line replaceable unit (LRU). These units are the modularized avionics systems equipment units (black boxes) that support communication, navigation, auto flight, in-flight entertainment, or other systems. If the connectors are not properly sealed, water will eventually enter the LRU through the connectors, resulting in premature failure or corrosion problems. The Department of Defense Corrosion Prevention and Mitigation Strategic Plan in November 2004 references one of the USAF sustainment priority projects as follows: "Improved avionics reliability through the use of corrosion-inhibiting lubricants" [69].

1.2.5.2 Use of corrosion inhibitive lubricants (CILs)

In the process of corrosion, thin and often invisible insulating films can form on the surfaces of electrical connectors by the reactions of natural environments with the elements used in the manufacture of commercial and military electrical connectors. Such films may represent a significant source of faults such as cannot duplicate (CND) and retest OK (RETOK) faults, the terms used in military avionics. USAF studies for the last twenty years have shown that application of certain *corrosion inhibitive lubricants* (CILs) or *corrosion preventative compounds* (CPCs) can substantially impact CND and RETOK rates, thus improving avionics reliability [70,71].

Corrosion between contacts may have materially contributed to several F-16 crashes. After the cause of the crash has been determined, corrosion-inhibiting lubricant spray, MIL-L-87177A Grade B (the primary component of MIL-L-87177A Grade B is polyalphaolephin (PAO), which is not flammable or hazardous), has been identified and is used annually in a preventative maintenance context. Treatment of F-16 electrical connectors with the MIL-L-87177A Grade B lubricant spray was remarkably effective. Conductivity of the tin-plated pins was fully restored and the CPC prevented continued corrosion, so much so that in a test at one base the aircraft so treated demonstrated a 16% improved mission capable (MC) rate.

In addition, millions of dollars saved by cost avoidances were documented by treating the aircraft and aircraft ground equipment (AGE) connectors [72]. The US Air Force has been using and testing MIL-L-87177A Grade B aerosol spray for corrosion control in electrical connectors for more than 10 years. As of March 2002, MIL-L-87177 Grade B has been added to the Air Force Tech Order/Manual 1-1-689, Avionics Cleaning and Corrosion Prevention/Control.

1.3 LUBRICANTS IN SPECIFIC ELECTRICAL CONTACT APPLICATIONS

1.3.1 Using Lubricants in Electronics

Lubricants are often used in the electronics industry, especially in conjunction with electrical connectors. They provide a number of benefits such as reducing the coefficient of friction, sealing plating pores, and reducing mating and un-mating forces. However, the use of lubricants may impair the processing of the connectors to the printed circuit board (PCB). A specific lubricants may improve performance of micro-electromechanical systems (MEMS).

1.3.1.1 Contamination of PCB electronic connectors by lubricants

Lubricant maybe applied to the connector assembly by different techniques: after plating or after final connector assembly. The latter is done by employing brush techniques, which allow for the "bleed over" of lubricants into areas where they are not needed. When the lubricant bleeds over to the lead or terminal areas where the electrical conductor terminates with PCB, it results in contamination becoming very problematic as the pitch and overall size of the metallic contact decreases.

A study [73] evaluated how various factors may effect the successful processing of connectors contaminated with lubricant in the lead region. These factors are lubricant manufacturer, lubricant viscosity, processing technology, flux type, and lead style. The conclusion of the study was that the probability of a processing problem for typical accidental lead contamination with any connector lubricant, any process, lead style, and flux is believed to be very low.

1.3.1.2 Effect of lubricants on PCB sliding contacts wear

Wear behavior of sliding contacts on printed circuit boards was studied in Ref. [74]. The grease was applied on flash gold contact pads in order to provide good oxidization protection and to allow contact functionality at temperatures down to −40°C. It was found out that depending on the PCB surface roughness, the mixture of grease and hard metallic wear debris could in some case accelerate the wear.

When using grease in a sliding contact, the roughness of the hardest partner has to be maintained below a certain value and kept under control. This value is determined experimentally with abrasion measurements during lifetime tests. The conclusion was that grease will improve wear only if the surface roughness of hardest contact surface is below a certain value, otherwise grease could even accelerate the wear out of the surface.

The grease can also provide negative effects by increasing the contact bouncing time and the minimum contact force due to an increased film resistance. The grease is an insulating layer in the contact interface, with a certain viscosity (base oil) and flow pressure. The presence of nonconductive particles (especially with PTFE based greases) is also a potential risk for the contact instability.

1.3.1.3 Lubrication of micro-electromechanical systems (MEMS)

Miniature micro-electromechanical systems often operate in extreme environmental conditions such as ultrahigh vacuum, high temperature, aero spatial, etc. Protecting MEMS against friction, wear, adhesion, and other phenomena that hinder performance and shorten operational life poses a significant challenge. To provide a proper functioning of such devices it is necessary to apply new lubrication materials in the field of low-level electrical contacts. Avoiding the traditional wet lubricating fluids would have many advantages.

A dry and chemically immobilized layer applied as a protective coating to the metallic surfaces of an electrical contact would be able to lubricate it, and at the same time preserve an electrical conduction. Dry films obtained by the electrochemical reduction of different *diazonium salts* have been studied in [75]. Gold surfaces used in electrical contacts have been coated with different diazonium salts and characterized by techniques such as ATR-IR and XPS and macroscopic friction tests. The coatings have shown outstanding wear resistance with low contact resistances.

Another development is reported in Ref. [76], which is presenting a promising lubricant—nanoparticles—to increase the durability of micro-electromechanical switches. In the study, gold nanoparticle lubrication increased average adhesion force by 44.51% and decreased average electrical-contact resistance by 76.40%.

The role of lubrication in reducing high friction in MEMS devices was studied in Ref. [76]. It was found that liquid lubrication may provide a solution to the problem of high friction and wear in MEMS. The methods of supplying lubricants to the sliding contacts in MEMS devices was developed in Ref. [77].

Excellent mechanical and electrical properties over traditional materials and coatings make graphene very attractive for a wide range of electromechanical applications ranging in sizes from nano/micro-scales (NEMS, MEMS) to macroscale (electrical contacts, sliding/rolling, rotating and bearings).

1.3.2 Graphene as a Lubricant and an Additive in Sliding Electrical Contacts

It was shown [78] that graphene demonstrates stable and low electrical resistance at the sliding contacts undergoing thousands of sliding passes regardless of the test environment (i.e., both in humid and dry conditions). Graphene was found to be a potential solid lubricant in sliding Ag-based electrical contacts [79].

Graphene can be easily and quickly deposited by evaporating a few droplets of a commercial graphene solution in air. The study showed that an addition of graphene reduced the friction coefficient in an Ag/Ag contact with a factor of ~10. A reduction in friction coefficient was also observed with other counter surfaces such as steel and W but the lifetime was strongly dependent on the materials combination. Ag/Ag contacts exhibited a significantly longer lifetime than steel/Ag and W/Ag contacts. The trend was explained by an increased affinity for metal–carbon bond formation.

It was found in Ref. [80] that the tribological performance of lithium grease could be significantly improved by the addition of graphene.

1.3.3 Challenge in Lubrication of Rolling and Sliding Electrical Contacts in Vacuum

Lubrication of the rolling and sliding electrical contacts in vacuum is a problem because of high coefficients of friction and excessive wear rates due to the absence of beneficial surface films. A study of the lubricants for such contacts in vacuum was published in 1968 [81]. It was found that graphite cannot be used in vacuum because its lubricating ability is dependent upon surface contamination.

The use of a good vacuum lubricant—molybdenum disulfide (MoS_2)—is questionable because of its high bulk resistivity and semiconducting characteristics. Niobium diselenide, which has a lower resistivity than MoS_2, is also a good vacuum lubricant although its wear rate is somewhat higher than that of MoS_2 when run against the same base material.

The use of high vapor pressure organic materials as electric contact lubricants in a space environment is questionable because of the possibility of undesirable polymer formation and the low radiation tolerance of such materials.

Composites containing dielectric materials would not be acceptable as an electrical-contact lubricant in vacuum because of the development of an insulating film between the two conducting surfaces resulting in an infinite electrical resistance of the contact.

Thin metallic films have been found to be a promising method of lubrication of sliding or rolling electrical contacts in vacuum, offering the greatest number of possibilities. The use of thin silver films as lubricants has resulted in long useful lifetimes under the severe conditions encountered in rotating anode X-ray tubes. The most pronounced disadvantage of this type of lubrication is that the useful life of the lubricant film is limited.

1.3.4 Graphite as a Lubricant for Monolithic Silver Brushes

Using graphite for lubrication of monolithic silver brushes required special attention because of high wear rates in high-power electrical motors. A new solution was suggested in Ref. [82] by loading a sacrificial solid lubricant, *graphite*, against the rotor to apply a thin, conductive transfer film to the interface of the sliding electrical contact in situ to reduce wear and friction while retaining high electrical efficiency.

These films may help in improving the operational life and efficiency of brush/rotor sliding electrical contacts. Decoupled in situ lubrication of monolithic silver brushes with graphite has demonstrated the ability to eliminate mechanical wear at low sliding velocities (1.6 m/s) and low current densities. High current densities, however, still present the problem of electrically induced wear of the brush contacts and significant Ohmic heating due to the thickness of the graphite transfer layer.

1.3.5 Lubrication of Electrical Contacts of Overhead Lines

Overheating of electrical contact is a very common phenomenon during the service of transmission and transformation equipment. It occurs at the connection joints of network and super-high-voltage equipment and super-current lines. The overheating defects are always a serious problem in an electric network; they not only increase wastage but also cause transformation accidents. With the increase of running load, the overheating defects of electrical contact become more frequent and serious. Electrical contacts of overhead lines are not usually lubricated.

However, the application of electrical joint compound to electrical contacts in transmission overhead lines was studied in Ref. [83]. This study demonstrated that electrical joint compound is one of the most effective materials to solve the overheating problem. When it was applied to aluminum-aluminum connections, the contact resistance was reduced by about 50% and the temperature of electrical connection decreased 20°C. The results show that electrical joint compound can reduce contact overheating and increase the reliability of electrical connection, since it also improves the corrosion resistance.

It was shown that the use of electrical joint compound on electrical contacts protects contact surfaces from corrosion, reduces friction and wear, and thereby reduces the contact resistance. Accordingly, electrical joint compound is an effective mean to solve the temperature rise, which not only can greatly enhance the joint reliability, but also plays a role in energy conservation. However, electrical joint compound must be carefully selected and subject to strict inspection before putting into service [84].

REFERENCES

1. Slade P.G. 1999. Introduction to contact tarnishing and corrosion, In *Electrical Contacts*, ed. Paul G. Slade, Marcel Dekker, Inc., New York, NY, Chapter 2: 89–112.
2. Timsit R.S. *Electrical Connector Lubricants.* http://www.timron-inc.com/newsletter.htm
3. Holm R. 1999. *Electric Contacts: Theory and Application*, 4th edition, Springer, New York: 483.
4. Van Dij P. 2002. Critical Aspects of Electrical Connector Contacts, January. http://www.pvdijk.com/images/21thiceccriticalaspects.pdf
5. Rudnick L. R. 2009. *Lubricant Additives: Chemistry and Applications*, Boca Raton: CRC Press/Taylor & Francis Group: 790.
6. Timsit R. S., Bock E. M., Corman N. E. 1997. Effect of surface reactivity of lubricants on the properties of aluminum electrical contacts, *Proc. 43rd Annual Holm Conference on Electrical Contacts*, Philadelphia, PA: 57–66.
7. Zhang J.G. 1994. The application and mechanism of lubricants on electrical contacts, *Proc. 40th Annual Holm Conference on Electrical Contacts*, Chicago, IL: 145–154.
8. Chudnovsky B. 2005. Lubrication of electrical contacts, *Proc. 51st Annual Holm Conference on Electrical Contacts*, Chicago, IL: 107–114.
9. Braunovic M. 1984. Evaluation of different types of contact aid compounds for aluminum to aluminum connectors and conductors, *Proc. International Conference on Electric Contacts Phenomena and 30th Annual Holm Conference on Electrical Contacts*, Chicago, IL: 97–104.
10. Abbott W.H. 1998. Performance of the gold-tin connector interface in a flight environment, *Proc. 44th Annual Holm Conference on Electrical Contacts*, Washington, DC: 141–150.
11. Timsit R.S. *Electrical Connector Lubricants.* http://www.timron-inc.com/newsletter.htm
12. Glossenbrenner E.W. 1999. Sliding contacts for instrumentation and control, In *Electrical Contacts*, ed. Paul G. Slade, Marcel Dekker, Inc.: 885–942.
13. *The Lubrication Engineers Manual*, 1971. Ed. by Charles A. Bailey, and J.S. Aarons, United States Steel, p. 460.
14. Lansdown A.R. 1996. In *Lubrication and Lubricant Selection, A Practical Guide*. UK Mechanical Engineering Publications, Pergamon Press, Oxford, UK, Section 1: 285.
15. Drozda T.J. and Wick C. 1983. *Fundamentals of Lubrication*. Tool and Manufacturing Engineers Handbook (TMEH Series), Society of Manufacturing Engineers, Dearborn, MI, Vol. 1, Machining, Chapter 4: 4-35–4-60.
16. *Lubricants and Lubrication*. 1992. ASM Handbook, Vol. 18, Ed. Peter J. Blau, ASM International, Materials Park, OH: 79–171.
17. Lubrication. 2010. SKF bearing maintenance handbook, Chapter 7, SKF Group Publications, November, 38.

18. Stevens C. 1995. Lubricant selection vital to maintenance solutions, *Plant Eng.*, 49(10): 64–68, August.

19. Antler M. 1984. Survey of contact fretting in electrical connectors, *Proc. International Conference on Electric Contacts Phenomena and 30th Annual Holm Conference on Electrical Contacts*, Chicago, IL: 3–22.

20. Capp P.O. and Williams D.W.M. 1984. Evaluation of friction and wear of new Palladium alloy inlays and other electrical contact surfaces, *Proc. International Conference on Electric Contacts Phenomena and 30th Annual Holm Conference on Electrical Contacts*, Chicago, IL: 410–416.

21. Gagnon D. and Braunovic M. 2002. High temperature lubricants for power connectors operating at extreme conditions, *Proc. 48th Annual Holm Conference on Electrical Contacts*, Orlando, FL: 273–282.

22. Sugimura K. and Nacae A. 1990. Lubricants for some plated contacts, *Proc. 36th Annual Holm Conference on Electrical Contacts*, Montreal, Canada: 417–424.

23. van Dijk P. 1992. Some effects of lubricants and corrosion inhibitors on electrical contacts, *AMP Journal of Technology*, 2: 56–62.

24. Antler M. 1999. Tribology of electronic connectors: Contact sliding wear, fretting, and lubrication, In *Electrical Contacts*, ed. Paul G. Slade, Marcel Dekker, Inc.: 403–432.

25. Aukland N., Hardee H., and Lees P. 2000. Sliding wear experiments on clad gold-nickel material systems lubricated with 6-Ring polyphenyl ether, *Proc. 46th Annual Holm Conference on Electrical Contacts*, Chicago, IL: 27–35.

26. Braunovic M. 1984. Evaluation of different types of contact aid compounds for aluminum to aluminum connectors and conductors, *Proc. International Conference on Electric Contacts Phenomena and 30th Annual Holm Conference on Electrical Contacts*, Chicago, IL: 97–104.

27. Abbott W. H. 1999. Contact corrosion, In *Electrical Contacts*, ed. Paul G. Slade, Marcel Dekker, Inc.: 113–154.

28. Antler M. Electronic connector contact lubricants: The polyether fluids, 1986. *Proc. 32nd Annual Holm Conference on Electrical Contacts*, Boston, MA: 35–44.

29. Zhang J.G., Mei C.H., and Wen X.M. 1989. Dust effects on various lubricated sliding contacts, *Proc. 35th Annual Holm Conference on Electrical Contacts*, Chicago, IL: 35–42.

30. Zhang J.G. 1994. The application and mechanism of lubricants on electrical contacts. *Proc. 40th Annual Holm Conference on Electrical Contacts*, Chicago, IL: 145–154.

31. Antler M. 1995. Corrosion control and lubrication of plated noble metal connector contacts, *Proc. 41st Annual Holm Conference on Electrical Contacts*, Montreal, Canada, pp. 83–96.

32. Basic connection principles, In *Burndy Reference:* O2–O5. http://portal.fciconnect.com/res/en/pdffiles/brochures/MC02_Section_O-Reference.pdf

33. *Engineering and Design—Lubricants and Hydraulic Fluids.* 1999. U.S. Army Corps of Engineers (USACE), Manual No. 1110–2–1424.

34. Braunovic M., Konchits V.V., and Myshkin N.K. 2006. Lubricated electrical contacts, *In Electrical Contacts, Fundamentals, Applications and Technology*, Section 9.3: 414–454, Boca Raton: CRC Press/Taylor & Francis Group.

35. Snyder D.R. December 2005/January 2006, Unearthing root causes, *MRO Today Magazine.* http://www.progressivedistributor.com/mro/archives/Uptime/UnearthingCausesD05J06.htm

36. Khonsari M. and Booser E.R. 2003. Predicting lube life—heat and contaminants are the biggest enemies of bearing grease and oil, *Machinery Lubrication Magazine,* September issue. http://www.machinerylubrication.com/Read/537/predict-oil-life

37. Khonsari Michael, Booser E.R. 2007. Low Temperature and Viscosity Limits, *Machinery Lubrication,* March. http://www.machinerylubrication.com/Read/1014/low-temperature-viscosity-limits

38. Kulwanoski G., Gaynes M., Smith A., Darrow R. 1991. Electrical contact failure mechanism relevant to electronic packages, *Proc. 37th Annual Holm Conference on Electrical Contacts,* Chicago, IL: 184–192.

39. Freitag W.O. 1976. Lubricants for separable contacts, *Proc. Holm Seminar on Electrical Contacts,* Illinois Institute of Technology, Chicago, IL: 57–63.

40. Abbott W. H. 1996. Field and laboratory studies of corrosion inhibiting lubricants for gold plated connectors, *Proc. 42nd Annual Holm Conference on Electrical Contacts,* Chicago, IL: 414–428.

41. Steenstrup R.V., Fiacco V.M., and Schultz L.K. 1982. A Comparative study of inhibited lubricants for dry circuit, sliding switches, *Proc. 28th Holm Conference,* IIT, Chicago, IL: 59–68.

42. McCarthy S.L., Carter R.O., and Weber W.H. 1997. Lubricant—induced corrosion in copper electrical contacts, *Proc. 43rd Annual Holm Conference on Electrical Contacts,* Philadelphia, PA: 115–120.

43. Klungtvedt K. 1996. A study of effects of silicate thickened lubricants on the performance of electrical contacts, *Proc. 42nd Annual Holm Conference on Electrical Contacts,* Chicago, IL: 262–268.

44. Witter G. J. and Leiper R. A. 1992. A study of contamination levels measurement techniques, testing methods, and switching results for silicon compounds on silver arcing contacts, *Proc. 38th Annual Holm Conference on Electrical Contacts,* Pittsburgh, PA: 173–180.

45. Pawlikowski G. T. 2006. Tyco Electronics HM-15 Lubricant Stability Projected Weight Loss, Technical Paper, Tyco Electronics Corporation (TE Connectivity Company). http://www.te.com/documentation/whitepapers/pdf/Tyco_Elec_HM-15_Wt_Loss_Stability.pdf

46. ASTM Standard D4048. 2016. *Standard Test Method for Detection of Copper Corrosion from Lubrication Grease,* ASTM International, West Conshohocken, PA, United States.

47. Wright J. 2008. Grease basics, *Machinery Lubrication,* May–June issue: 50–52. http://www.machinerylubrication.com/Read/1352/grease-basics

48. *Lubrication Manual: Lubrication Instructions for Major Manufacturers' Medium and Low Voltage Circuit Breakers,* 2008. 6th Edition, Schneider Electric/Square D: 84.

49. Richard D.M. 1996. Lubrication products and practices in circuit-breaker maintenance, *Power Supply World:* 69–72.

50. Crino A.D. 1996. The effects of exposed contact lubricants on the temperature rise of outdoor disconnect switches. *Proc. 63rd Annual International Conference of Doble Clients,* Boston, MA: 4-2.1–4-2.5.

51. Chudnovsky B. 2001–2002. Lubrication practices in the electrical Industry: The key to reliability of circuit breakers, *NETA World*, Winter: 55–59.
52. *Lubrication Guide of the Doble Circuit Breaker Committee*, 1995. Doble Engineering Company, Watertown, MA, Rev C: 27.
53. Chudnovsky B. 2003. Lubrication practices can make or break circuit breaker reliability, *Lubrication and Fluid Power*, 4(4): 23–26, November/December issue.
54. Chudnovsky B. 2004. Ensuring protection and reliability of load interrupter switches, *Plant Safety and Maintenance*, 4(2): 31.
55. Chudnovsky B. 2008. Smooth breakers. *Plant Services*, November issue. http://www.plantservices.com/articles/2008/246.html
56. *Lubrication and Circuit Breakers*, 2001. Square D Data Bulletin # 0600DB0109. https://download.schneider-electric.com/files?p_enDocType=Data+Bulletin&p_File_Name=0600DB0109.pdf&p_Doc_Ref=0600DB0109
57. *Preventative Maintenance for HVLTM Load Interrupter Switches*, 2002. Square D Data Bulletin #6040DB0201. https://download.schneider-electric.com/files?p_enDocType=Data+Bulletin&p_File_Name=6040DB0201.pdf&p_Doc_Ref=6040DB0201
58. Anderson A. F. 2007. Reliability in Electromagnetic Systems: The role of electrical contact resistance in maintaining automobile speed control system integrity, Delivered at the IET Colloquium on Reliability in Electromagnetic Systems, Paris, 24th May. Published in Conference Digest: PEZ07827 (07/11827). http://www.theiet.org/publishing/conf-pubs/semdig07.cfm
59. Joaquim M. E., Adams T. Stationary Lubricants increase connector reliability, Santovac Fluids Inc., Technical Paper. www.chemassociates.com/products/findett/PPEs_Swedish_Cell.pdf
60. Bob Mroczkowski. 2014. Connector contact lubricants, ConnectorSupplier.com. http://www.connectorsupplier.com/connector-contact-lubricants/
61. Synthetic Lubes Protect Electrical Connections, Reduce Warranty Claims, NYE Synthetic Lubricants: 9. www.nyelubricants.com/_pdf/connector_tech_overview.pdf
62. Keeping contact: specialty lubricants for electrical contacts. 2012. Kluber Lubrication. Bulletin B053001002, Edition 11-2012, München, Germany: 1-11. http://pdf.directindustry.com/pdf/kluber-lubrication/keeping-contact-lubricant/12008-367115.html
63. MIL-DTL-87177B, 2014. Detail Specification: Lubricant, Corrosion Preventive Compound, Water Displacing, Synthetic (23-SEP-2014), Superseding MIL-L-87177A.
64. Hamid S. 2006. Using Lubricants to Avoid Failures in Medical Electronic Connectors, *Medical Electronic Design*, October issue. http://www.medicalelectronicsdesign.com/article/using-lubricants-avoid-failures-medical-electronic-connectors
65. *Synthetics, Mineral Oils, and Bio-Based Lubricants: Chemistry and Technology*, 2013. Edited by Rudnick L. R., Second Edition, Boca Raton: CRC Press/Taylor & Francis Group: 194.
66. Mraz S., Cichoski B. 2004. Giving electrical connectors the slip, *Machine Design*, February. http://machinedesign.com/archive/giving-electrical-connectors-slip

67. Automotive Electrical Contact Lubricants, Syn-Tech, Ltd. Technical Sheet. http://www.syn-techlube.com/userfiles/pdf/tech-sheets/Syn-Tech_Auto_Electrical_web.pdf

68. Sabatini N. A. 2001. Flight Standards Service, "Inspection, Prevention, Control, and Repair of Corrosion on Avionic Equipment", US D.O.T F.A.A Advisory Circular.

69. Report to Congress, Department of Defense Status Update on Efforts to Reduce Corrosion and the Effects of Corrosion on the Military Equipment and Infrastructure of the Department of Defense. 2005. Undersecretary of Defense AT&L. http://www.nstcenter.com/docs/PDFs/TechResourcesReportToCongress.pdf

70. Abbott W.H., Hernandez H., Kinzie R, Siejke S. 2002. Effects of corrosion inhibiting lubricants on electronics reliability, *Proc. of the 21st Digital Avionics Systems Conference*, v. 2: 12D2-1–12D2-13.

71. Dotson S. L. Effects of corrosion inhibitive lubricants on electronics reliability: 1–14. http://correxllc.com/resources/pdf/Corrosion_inhibitive_lubricants_Scott_Dobson.pdf

72. Horne D. H. 2000. Catastrophic Uncommanded Closures of Engine Feedline Fuel valve from Corroded Electrical Connectors, Paper# 00719, NACE. www.lektrotech.com

73. Meredith K. 2016. *Effects of Lubrication on Connector Processing*, Samtec, Inc. http://suddendocs.samtec.com/notesandwhitepapers/effects-of-lubrication-on-connector-processing.pdf

74. Le Solleu J.-P. 2010. Sliding contacts on printed circuit boards and wear behavior, European Physical *Journal: Applied Physics, EDP Sciences*, 50 (1): 12902. https://hal.archives-ouvertes.fr/hal-00581869/document

75. Ko S-D., Seo M-H., Yoon Y.-H., et al. 2017. Investigation of the Nanoparticle Electrical Contact Lubrication in MEMS Switches, *Journal of Microelectromechanical Systems*, Volume 26, Issue 6: 1417–1427, October.

76. Ku I.S.Y., Choo J.H., Holmes A.S., Spikes H.A. MEMS Lubrication, Engineering and Physical Sciences Research Council. www3.imperial.ac.uk/pls/portallive/docs/1/9695696.pdf

77. Leong J. Y., Zhang J., Sinha S. K., et al. 2015. Confining Liquids on Silicon Surfaces to Lubricate MEMS, *Tribology Letters*, vol. 59, issue 15: 11.

78. Berman D., Erdemir A., Sumant A.V. 2014. Graphene as a protective coating and superior lubricant for electrical contacts, *Applied Physics Letters*, vol. 105, issue 23. http://aip.scitation.org/doi/am-pdf/10.1063/1.4903933

79. Mao F., Wiklund U., Andersson A.M, Jansson U. 2015. Graphene as a lubricant on Ag for electrical contact applications, *J Mater Sci*, v. 50: 6518–6525. https://link.springer.com/content/pdf/10.1007%2Fs10853-015-9212-9.pdf

80. Jing W., Xiaochuan G., Yan H., et al. 2017. Tribological Characteristics of Graphene as Lithium Grease Additive, *China Petroleum Processing and Petrochemical Technology*, Lubrication Research, Vol. 19, No. 1: 46–54. www.chinarefining.com/EN/article/downloadArticleFile.do?attachType=PDF&id

81. John S. Przybyszewski. 1968. A Review of Lubrication of Sliding and Rolling-Element Electrical Contacts in Vacuum, NASA Technical Note TN-4476, 33. https://ntrs.nasa.gov/archive/nasa/casi.ntrs.nasa.gov/19680012510.pdf

82. Bares J.A., Argibay N., Dickrell P.L, Sawyer W.G. 2009. In situ graphite lubrication of metallic sliding electrical contacts, *Wear:* 1–8.
83. Wang G., Zhao Y., Wu L., Zhigang K. 2015. Influence of Electrical Joint Compound on Electrical Contact Overheating of Overhead Line, 2nd International Workshop on Materials Engineering and Computer Sciences (IWMECS): 128–132.
84. Wang G, Wu L., Li G., Zhang G., Guanglei Z., 2014. Protection and Practice of Overheating of Electrical Contact of Power Transmission and Transformation Equipment, *International Conference on Mechatronics, Electronic,* Industrial and Control Engineering (MEIC): 475–478.

Lubrication of Mechanical Components in Electrical Equipment

2

2.1 ROLE OF MECHANISM LUBRICATION IN CIRCUIT BREAKING EQUIPMENT

Circuit breaker mechanisms have multiple components that must properly operate for the breaker to fulfill its function. If one component does not operate correctly, the circuit breaker may not operate correctly or fail. It has been found that a significant portion of circuit breaker operations failures are caused by deteriorated or inadequate lubrication of breaker mechanism components, such as bearing rollers and sliding surfaces.

2.1.1 High-Voltage Circuit Breakers

When a lubrication-related failure occurs in high-voltage circuit breakers (HVCBs), backup relays, adjacent breakers, and associated components operate to clear the problem. When an HVCB fails the event can rapidly cascade, affecting the electrical grid and potentially causing equipment damage and outages, with significant customer service and financial impact [1].

High-voltage circuit breakers require very fast operation of complex mechanical assemblies for successful operation. Consequently, various critical components of these mechanisms, such as bearings, must be properly lubricated to assure correct motion. Proper lubrication of circuit breaker components reduces friction and wear between moving surfaces, typically metal on metal; reduces or prevents corrosion and rust; repels many contaminants; and prevents heat buildup.

Understanding the importance of lubrication on critical components in circuit breaker mechanisms is more complex than most other lubrication applications. Circuit breaker mechanisms may stay dormant for long periods. Lubricated surfaces in a mechanism, such as bearings or sliding surfaces, do not perform full and continuous rotation as they are typically designed.

Circuit breaker manufacturers usually specify one lubricant for a breaker type regardless of operating environment. Utility crews are not always aware of grease compatibility issues or proper application techniques. Commercial greases are formulated for industrial applications where operation is usually continuous and maintenance intervals more frequent than typical utility practice. Intrinsic factors such as significant temperature cycling and harsh environmental conditions may accelerate aging of the lubricants.

Aged or improperly specified or applied greases can slow mechanism operation beyond acceptable limits. Therefore, the choice of lubricants can affect the reliable operation of a breaker. Using incompatible or unapproved lubricants can lead to unacceptable performance or premature failure [1].

2.1.2 Low- and Medium-Voltage Circuit Breakers

There are a number of recommendations in regards to the lubrication procedures and lubrication products applied to low-voltage circuit breakers (LVCBs) and medium-voltage circuit breakers (MVCBs). The same recommendations are applicable to mechanical parts of switches and contactors: moving surfaces and bearings [2].

If grease on the mechanical components appears to be discolored it generally means that *oxidation* is taking place and the problem must be addressed. The old grease must be removed, which may require partial disassembling of the mechanism.

There are several chemicals that may be used for removing the old grease. In the past, alcohol, kerosene, and other flammable products were recommended, which introduced an unacceptable hazard. The nonflammable chlorinated products (trichloroethylene and its variations) were a welcome solution, especially around electrical equipment. However, these have environmental and toxic issues that eliminate their use in some areas. The citrus cleaners are not toxic and are environmentally friendly, but they leave a residue which is conductive and should be cleaned off. When choosing a proper cleaner it is necessary for the chemicals be compatible with plastics. The parts should be dry before applying new grease.

One of the most damaging technique for cleaning is the use of a spray penetrating oil, such as WD-40, sprayed into the operating mechanism. It contains a small percentage of lubrication oil, but the rest of its composition is a solvents, which flush out any remaining lubricant and does not provide any lasting lubrication. WD-40 and other liquid penetrants do not act as lubricants and should not be used as lubricants. They act as solvents and will only temporarily free up sticky interfaces [3].

If during the maintenance the contaminated grease is found inside a mechanism, it should be completely refurbished. Before replacing the bearings the old grease should be completely washed out and then the bearings packed with the new grease to avoid the incompatibility problem (see Section 2.2.4). All lubricating maintenance procedures should be performed accordingly to the manufacturer recommendation, including which point to lubricate and which lubrication products to use.

2.1.3 Types of Lubricants Used in Mechanisms

There are three major types of the lubricants: oils (fluid lubricants), greases (soft substances), and solid lubricants. *Oils* covers a broad class of fluid lubricants, each of which has particular physical properties and characteristics.

Basically, *petroleum or mineral* oils are made from naphthenic or paraffinic crude oil. Naphthenic oils contain little wax and their low pour point makes them good lubricants for most applications. Paraffinic oils, on the other hand, are very waxy. Lubricants made from them are used mainly in hydraulic equipment and other machinery.

> *Synthetic oils* are being used for instrument bearings, hydraulics, air compressors, gas and steam turbines and many other applications. *Silicones* are not true oils; their principal advantages are excellent viscosity, temperature characteristics, good resistance to oxidation, and a wide operating temperature range.
>
> *Grease* consists of a thickening agent, oil or synthetic fluid, and additives. Various thickeners provide different properties of grease [4–6]. Greases are usually ranked by their consistency on a scale set up by the National Lubricating Grease Institute (NLGI). Consistency

or grade numbers range from 000 to 6, corresponding to specified ranges of penetration numbers.

Synthetic lubricants have esters, synthetic hydrocarbons (SHC), poly-glycols, silicones, and so on as a base stock. Synthetic greases provide a cost-effective, lifetime lubrication for bearings and other moving parts, and for gaskets and seals. High thermal stability and chemical inertness make them useful for aerospace, electrical, automotive, and other high-tech or industrial applications. Synthetic lubricants usually keep their lubrication characteristics at temperatures in the range of −73°C to 290°C (−100°F to 550°F), maintain good viscosity over a wide temperature range, and are nonflammable with no autogeneous ignition, flash, or fire point up to 650°C (1,200°F).

Synthetic lubricants are available in a variety of forms, from light oils to thick greases. Synthetic lubes [7] have a longer product life, are more inert, and generate less waste than petroleum materials. They are capable to perform in a wide range of temperatures, from extremely low to high, and certain classifications are friendly to elastomers, seals, and O-rings that come in contact with the lubricant.

Solid Lubricants are usually fine powders, such as molybdenum disulfide, graphite, and Teflon (polytetrafluoroethylene or PTFE). They can be used as additives in greases or dispersions, as dry film bonded lubricants, or alone. Lubricating solids can provide longer-term lubrication than unfortified oils and greases because of their ability to form burnished films on surfaces and stay longer.

Silicones are very stable and very inert lubricants that provide wider operating temperature ranges than nonsilicone synthetic lubes. *Fluorosilicones* have some of the advantages of other silicones, such as dimethyl and phenylmethyl silicones, plus higher resistance to harsh environments and the ability to carry heavy bearing loads.

2.1.4 Thickeners in Greases Used in Mechanisms

Greases are manufactured by combining three essential components: base oil, thickener, and additives. Base oil (80% to 90%) provides lubrication and is a temperature limiting element. Thickeners (10% to 20%) provide consistency, keep the oil where needed, and act as temperature sensitive elements. Additives (3% to 5%): provide valuable properties such as anti-wear, anti-corrosion, and anti-oxidation; Among them are viscosity modifiers, detergents, dispersants and many other chemicals. Additional information is given in Addendum 1: Grease Composition and Properties.

The role of a thickener is to give grease its characteristic consistency and the thickener is sometimes thought of as a "three-dimensional fibrous network" or a "sponge" that holds the oil in place. Lubrication function has been compared to that of a sponge (thickener) gradually releasing its liquid over a period of time. There are many different classes of materials used as thickeners in grease manufacturing: simple and complex metal soaps and inorganic and organic non-soaps. Thickeners used in circuit breaker greases can be soap or non-soap.

> *Soap Thickeners.* Soap thickeners can be combined with salts to raise the grease dropping point (the temperature at which the grease liquefies). Common soap thickeners include lithium salt of hydrogenated castor oil fatty acid, calcium stearate (hydrous), calcium stearate (anhydrous), lithium complex, aluminum complex, and calcium complex.

Most used in industrial and manufacturing environments are lithium and lithium complex greases, which have strong properties in many categories. These greases have high dropping points and excellent load-carrying abilities, excellent long-term work stability, strong high-temperature characteristics, and acceptable wash- and corrosion-resistance capabilities. Lithium-soap greases have good freshwater resistance, but poor saltwater resistance.

> *Non-soap Thickeners.* Some lubricating greases are manufactured with organoclays, polyurea compounds, silica aerogel, and polytetrafluoroethylene (PTFE). A thickener's ability to resist water is also an important consideration. Even in devices that are generally protected from the environment, humidity can condense inside them and displace the grease, or it can become entrained and accelerate corrosion. Clay and PTFE are generally used for wet applications. PTFE also lowers friction, especially on plastic components.
>
> *Polyurea* greases are widely used in ball bearing applications. These greases contain little to no heavy metals and have favorable high-temperature performance resulting in very good oxidation-resistance, they provide good work stability, wash- and corrosion-resistance. Some polyureas have a low level of compatibility with other soap and non-soap greases, including other polyureas.
>
> *Bentonite* (type of clay) products are created by the direct addition of the thickener to the base and additive mixture. These products require significant milling to assure uniformity. Bentonite greases are incompatible with most other grease types. Some clay-thickened (bentonite) greases may have very high melting points,

with dropping points noted on the product data sheets as 500°C or greater; some data sheets say that there is no dropping point.

For these non-melting products, the lubricating oil burns off at high temperatures, leaving behind hydrocarbon and thickener residues. Other base oils may evaporate before the clay material becomes hot enough to melt. This is both a strength and weakness. When used for extended periods of time at elevated temperatures, bentonite grease residues may cause a filling of the housing that can make long-term relubrication difficult.

2.1.5 Greases for Industrial Circuit Breakers

Proper grease selection is essential for the electrical industry to minimize friction and wear between the components of circuit breakers under mechanical contact. Greases are used as lubricants due to their ability to stay in the original position where they were applied, avoiding relubrication, particularly for areas where it is no longer possible.

Greases should provide adequate lubrication to reduce friction and prevent harmful wear on components, protect the mechanism against oxidation and corrosion, act as a sealant preventing the passage of contaminating particles, maintain its viscosity regardless of stress or temperature during usage, avoid dripping, not over solidify to avoid undue resistance to motion in cold environments, and tolerate certain levels of contamination, such as humidity, without losing their characteristics significantly [8].

The choice of the best lubricants should be based on their anti-wear (AW) and extreme pressure (EP) characteristics. Presented in Table 2.1 are the results of the study [9] of several commercially available lubricating greases with different base oil, thickener, viscosity, and temperature range properties. The goal of the study was to determine the tribological properties of wear, load-carrying capacity, and pressure loss limit of greases. If the property was not supplied in the grease datasheet it is denoted as NA (not available). Highlighted in the table are the greases proven best in the following categories: 1AW—best in anti-wear capability, 1D—best in durability; 1LC—best in load-loading capacity; 2LC—second best in load-loading capacity; 3LC—third best in load-loading capacity.

Test results show that lithium-thickened PAO greases with a viscosity of 220 cSt have the best anti-wear tribological properties, meaning that for mechanical applications in industrial breakers viscosity is an important factor to consider together with working temperature range and the kind of thickener.

TABLE 2.1 Properties of greases selected for the study

GREASE #	BASE OIL, THICKENER; NLGI GRADE	BASE-OIL VISCOSITY (cSt)	WORKING T RANGE (°C, °F)	DROP POINT (°C)	FLASH POINT (°C)	POUR POINT (°C)	PENETRATION, UNWORKED/ WORKED	TEST RESULTS
1	PAO, Clay; 1.5	30	N/A	307	NA	NA	NA/293	
2	PAO, Li; 1.5	32.6	−54 to 125 (−65 to 257)	NA	250	−62	297/306	
3	PAO, Li; 2	32.6	−54 to 125 (−65 to 257)	NA	235	−62	274/277	
4	PFPE, Silica; NA	510	−25 to 250 (−13 to 482)	NA	None	−20	NA/NA	1LC
5	Ester, Clay; 2	54	−40 to 150 (−40 to 302)	NA	304	−54	248/267	
6	PAO, Li; 2	220	−25 to 140 (−13 to 284)	280	NA	−8	NA/280	1AW, 1D, 2LC
7	PAO, Li; 2	220	−34 to 177 (−29 to 350)	271	NA	NA	265/295	1AW, 3LC

2.1.6 Mechanism Bearings Failures Caused by Lubrication

Lubrication is necessary for a proper bearing function, but it also has disadvantages. A lubricated surface may retain more dust, the wiping motion may get less effective, and lubricant may enable sliding micromotion where this would not take place without lubrication. Lubricant can polymerize and form insulating films. It can form a varnish and can creep to places where it is not wanted. Multiple types of bearing failures are caused by lubricants.

According to a major ball bearing company, 54% of bearing failures are lubrication-related. In a study by MIT, it was estimated that approximately $240 billion is lost annually (across US industries) due to poor lubrication. Improper bearing lubrication or relubrication accounts for up to 40% to 50% of machine failures [10] which altogether lead to 40%–50% of such failures. There are numerous causes for lubricant failure, among them insufficient lubricant quantity or viscosity and deterioration due to prolonged service without replenishment. The lubricants also fail when exposed to excessive temperatures or being contaminated with foreign matter [11].

Inadequate lubrication can create a wide range of damage conditions. Damage happens when there is not a sufficient amount of bearing lubricant to separate the rolling and sliding contact surfaces during service. It is important that the right lubricant amount, type, grade, supply system, viscosity, and additives be properly engineered for each bearing system. Selection of lubricants should be based on history, loading, and speed, as well as sealing systems, service conditions, and expected life [12].

Without proper consideration of these factors, performance of the bearings may be lower than expected. The damage caused by inadequate lubrication varies greatly in both appearance and performance. Depending on the level of damage, it may range from very light heat discoloration to total bearing lockup with extreme metal flow. The levels of bearing damage caused by inadequate lubrication may progress from discoloration to scoring and peeling to excessive roller end heat to a total bearing lockup.

Lubrication with grease when conditions dictate the use of static or circulating oil also may lead to bearing failure. Oil is the preferred lubricant when speed or operating conditions preclude the use of grease or where heat must be transferred from the bearing. Often oil is used to meet the operating requirements of other components such as seals and gears.

If for a particular application an *incorrect grease base* is chosen, the bearing may fail. Grease selection should vary with the application. Factors to consider include hardness (consistency), stability (ability to retain consistency), and water resistance (emulsification). However, grease is an oil suspended in a base or carrier, and when these bases are exposed to moisture or heat, they can turn into soap or carbon ash. Therefore, it may be necessary to use synthetic additives to prevent deterioration of the base.

Preventing bearing failure requires a correct choice of lubricant viscosity, which is just as important as oil quantity. Required viscosity depends on operating temperature. Inadequate lubricant viscosity appears as a highly glazed or glossy surface. As damage progresses, the surface appears frosty and eventually spalls. This type of spalling is comparatively fine grained compared to the more coarsely grained pattern produced by fatigue failure [13].

A bearing may fail if it is *overlubricated* [14]. Overlubrication may cause a rapid rise in temperature, particularly at high speeds, because the rolling elements have to push the grease out of the way. This leads to churning in the grease, which produces heat. This churning action will eventually bleed the base oil from the grease and all that will be left to lubricate the bearing is a thickener system with little or no lubricating properties. The heat generated from the churning and insufficient lubricating oil will begin to harden the grease.

This will prevent any new grease added to the bearing from reaching the rolling elements. The end result is bearing failure and equipment downtime. Adding more grease only worsens the problem, adding the risk of blowing out a seal. Ironically, an attempt to sufficiently lubricate a bearing by giving it several extra pumps from a grease gun will eventually result in its failure due to under-lubrication.

2.1.6.1 Cake-lock bearing failures

The tendency of grease to cake and dry out is one of the main disadvantages of using grease over oil. When oil is drained from the grease, over time the hardened byproduct builds inside the cavities of operating bearings and gears. This process may take years to develop, while in other cases the condition can reach catastrophic levels in just a few weeks.

As grease becomes dry, frictional forces escalate, causing concentrated heat within the bearing. The elevated heat continues to force more rapid and complete drying of the grease. Eventually the grease can reach a consistency ranging from hard putty to sandstone, depending on the thickener type and degradation conditions [15].

Multiple conditions may lead to grease dry out. It could be gross contamination by dust, dirt, fly ash, etc. Dry contaminants can thicken the grease, as well as incompatibility of greases, when accidental mixing of greases may result in accelerated de-gelling and oil separation. When grease is made from low-viscosity base and oils having high-temperature volatility, oil can boil out of the thickener matrix at sufficiently high temperatures, causing hardening of the grease over time. This process leads to lower oil-to-thickener ratio and higher base-oil viscosity.

As the thickener dries out, it becomes immobile and jams the flow path and even mechanical motion. As more grease enters and oil exits, the more the logjam builds. Eventually, a rigid binding condition develops from the stiff, high-density cake. The cake-lock condition leads to thermal excursion and bearing failure. Bearings lubricated with centralized or single-point lubrication are perhaps most at risk for cake-lock. In centralized lube systems, cake blockage can occur in the lines between the pump and the dosage injectors. When soft grease is being fed into the bearing but only oil emerges from the exhaust port, there is a cake-lock failure in progress. The thickener is being log-jammed in the bearing cavity.

2.1.6.2 Hardened and oxidized lubricant in bearings

In ball bearings, the continuous presence of a very thin—millionths of an inch—film of lubricant is required between balls and races and between the cage, bearing rings, and balls. Failures are typically caused by restricted lubricant flow or excessive temperatures that degrade the lubricant's properties. Discolored (blue/brown) ball tracks and balls are symptoms of lubricant failure [16]. Excessive wear of balls, ring, and cages will follow, resulting in overheating and subsequent catastrophic failure. In rolling and ball bearings, improper lubrication may result in bearing fracture and seizure, formation of cracks, and wear on outer/inner rings and cage; if the grease has too high viscosity, it may cause scratches on raceway surface [17].

In addition, if a bearing has insufficient lubrication, or if the lubricant has lost its lubricating properties, an oil film with sufficient load-carrying capacity cannot form. The result is metal-to-metal contact between rolling elements and raceways, leading to adhesive wear. If the grease is stiff or caked and changed in color, it indicates lubrication failure. The original color will usually turn to a dark shade or jet black. The grease will have an odor of burnt petroleum oil. Lubricity will be lost as a result of lack of oil. In cases of lithium base greases, the residue appears like a glossy, brittle varnish that will shatter when probed with a sharp instrument.

One can observe strong oxidation and hardening of the grease that occurs following high-temperature stress, which is produced through electrical grounding (arcing). Loss of lubricant health produces mixed friction and wear in the roller contact area. The fact that a bearing cannot be easily relubricated from the outside plays a crucial role in eventual element failure. If the hardened and oxidized lubricant already presents in the bearing, the newly added grease cannot displace it. It makes an exchange of grease impossible even with normal relubrication intervals; bearing failure is inevitable [18].

2.1.6.3 Lubricant water contamination in bearings

Moisture enters bearing systems in several different ways, resulting in dissolved, suspended, or free water in the lubricant. Both dissolved and suspended water can promote rapid oxidation of the lubricant's additives and base stock resulting in diminished lubricant performance. As little as 1% water in grease can have a significant impact on bearing life.

Many different moisture-induced wear and corrosion processes are common in both rolling element and journal bearings. Rolling element bearings may experience reduced fatigue life due to hydrogen embrittlement caused by water penetrated bearing surfaces. A tapered roller bearing cone and rollers and a ball bearing outer race and balls show rusting with pitting and corrosion from moisture/water exposure. This condition is referred to as *etching* [19]. Etching is most often caused by condensate collecting in the bearing housing from temperature changes. Moisture or water also can get in through damaged, worn, or inadequate seals. The hydrolysis reactions between the lubricant and suspended water result in the liberation of free sulfur, which produces hydrogen sulfide and sulfuric acid from water-induced lubricant degradation. Water etching is a common type of corrosion occurring on bearing surfaces and their raceways.

Improperly washing and drying bearings when they are removed for inspection can also cause considerable damage. Moreover, moisture can contaminate the lubricant and condense on the gear surfaces forming sludge, corrosion, and micropitting. The best defense against moisture contamination is a three-step, proactive maintenance strategy called Target-Exclusion-Detection (TED). Only when lower moisture levels are consistently stabilized can the life extension of lubricants and bearings be effectively achieved.

2.2 FACTORS CAUSING LUBRICATION FAILURES IN MECHANISMS

The lubricating effectiveness of grease is limited by both physical and chemical deteriorations caused by shear stresses, pressure, and the severity of operating conditions, particularly temperature. Degraded grease becomes inefficient and eventually loses its lubrication capacity such that it can adversely affect the performance and functionality of mechanical parts. Grease degradation is categorized into physical and chemical degradation.

Physical degradation covers all the physical changes in the grease during usage. This degradation category includes the integral parts of mechanical degradation (breaking of grease thickener's structure), increased base-oil separation, base-oil evaporation, and also increased contamination.

Chemical degradation covers all the chemical reactions inside the grease that affect grease lubricating life including additive depletion (mostly antioxidant), base-oil oxidation, and thickener oxidation.

Physical degradation is more sensitive to applied force (shear stress) and velocity (shear rate), and chemical degradation is more sensitive to operating temperature [20].

2.2.1 Oxidation and Gelling

The lubricating properties of greases are reduced or become non-functional for many reasons depending on the type oil and thickeners, application method and thickness of grease applied, environmental factors such as high and low temperature, chemical exposure, and length of time in service.

Oxidation. The primary failure mechanism of hydrocarbon (mineral) greases is oxidation of the base oil in the grease. Exposure of lubricant to atmospheric oxygen forms gums, resins, and acidic products with viscosity increase. When the grease oxidizes, it usually darkens; there is a buildup of acidic oxidation products. These products can have a destructive effect on the thickener, causing softening, oil bleeding, and leakage. Because grease does not conduct heat easily, severe oxidation can begin at a hot point and spread slowly through the grease. This produces carbonization and progressive hardening or crust formation [21]. The sequence of oxidation is as follows: oil combination with oxygen, evaporation of low molecular weight oxidates, and depletion of additives as oxidation continues. The additives are generally used to buffer the effects of oxygen exposure and are used up as exposure to oxygen continues. Oxidation accelerates rapidly after the additives are depleted. The end result is usually sludge and/ or a yellowish varnish deposit on equipment parts. When the sludge or varnish occurs, it is the end of the lubricating properties of the oil or grease. To protect the grease from contact with oxygen it is recommended that lubricants containing antioxidant additive be used.

Gelling. Fluorosilicone Grease (FSi) oil is highly inert, therefore oxidation is not a usual failure mechanism for FSi grease. Gelation with viscosity increase is an age and temperature related fluorosilicone oil failure mode. Gelation makes the oil molecules larger and viscosity higher. Lubricating properties of FSi grease do not seem to be impaired as this gelling occurs. The bearing grease in the study [1] showed various degrees of gelling. The thickening or gelling of the oil was not sufficient to slow the operation of the mechanisms. This study recommends using fluorosilicone greases with PTFE thickener when lubricating bearings or sliding surfaces in circuit breaker mechanism applications despite possible gelling of base oil.

2.2.2 Environmental Factors

Grease is exposed to many environmental factors that cause contamination and oxidation of the grease. These factors are atmospheric oxygen, dust and dirt, high and low temperatures, and light. *Dirt and dust* in the grease increase the rate of wear between bearing surfaces, which also necessitates keeping the lubricants covered or in containers [22].

Exposure of the grease to *high temperatures* increases the rate of deterioration; forms gums, resins, and acidic products; and increases oil separation from greases, while the exposure of water-containing materials (e.g., cutting oils and certain fire resisting hydraulic oils) to low temperatures causes water to separate out. Therefore, it is recommended to keep the temperature of lubricant storage not higher than 20°C (68°F) and above the freezing point. *Light* promotes the formation of gums, resins, and acidity in the lubricants; to protect the materials from exposure to light, they should be stored in metal or opaque containers.

With increasing time in service and temperature fluctuations, lubricants may degrade by oxidation or polymerization, forming insulating films. Oxidation occurs in all oils and greases that are in contact with air, including stored lubricants. The base oil and additive combination affects the rate of oxidation, and the presence of a thickener in grease can increase the degradation rate.

If the lubricant was badly degraded or reacted chemically to *environmental contaminants*, deposits in the form of gum, resin, varnish, sludge, or other deposits could be found in places where the lubricant was applied. Elemental analysis of these residues may help to identify the sources of deterioration and to determine if they were degraded grease additives or thickeners or environmental contaminants.

2.2.3 Contaminants and Corrosives

To properly function, grease should be free of contaminants. Contaminated grease will reduce bearing performance and lifetime. Contamination may come from the outside, introduced by poor sealing, dirty grease guns, or poor bearing mounting methods. Contaminants can be found in grease used for lubrication of any moving parts of electrical equipment, particularly in mechanical parts.

Grease contamination can be caused by various natural sources such as sand, water, dust, fibers, steam flow, etc. Lubricant contamination may be a result of oil leaking from neighboring systems or introduction of wrong or incompatible grease. If the grease reaches end-of-life it deteriorates, forming carbonized particles that stick to surfaces promoting friction. The grease used in the bearing is contaminated with bearing wear material [23].

More than 20 different elements can be found in used grease. Information about the elemental composition of the used grease may help to evaluate degree of wear, contamination, and the condition of the additives and thickeners. These elements include wear metals such as iron, chromium, tin, copper, lead, nickel, aluminum, molybdenum, and zinc. Contamination from the environment may include such elements as silicon, calcium, sodium, potassium, and aluminum. Decomposition of additives or thickeners may add magnesium, calcium, phosphorous, zinc, barium, silicon, aluminum, molybdenum, and boron to the grease composition [22].

Besides solid contaminants, which can be identified by the presence of silicon, calcium, or aluminum, water is a type of contamination that is often the cause of corrosion. Depending on the grease type and application, the water content in the grease should not exceed the recommended values. Too much water in grease can produce a variety of adverse effects, including corrosion on bearing metals, increased oxidation of the base oil, softening of the grease, and water washout of the grease.

Although contaminants are sometimes difficult to detect, they often cause lubrication failure in circuit breakers. Dirt, sand, and water are the most common contaminants, but acid and other corrosives also can deteriorate the lubricant (grease and oil). They can dilute the oil film, reducing viscosity, and they can corrode metal surfaces, disrupting the lubrication film, causing erosion, and creating thousands of abrasive particles. Moisture contamination increases chemical wear by rusting iron and steel surfaces; it also increases the corrosive strength of the acids attacking copper and lead surfaces [22].

Solid particulate contaminants introduce the most damage when they come into contact with the surfaces of rotating parts (bearings). Depending on particle size, they may be seen or felt as grittiness in the lubricant. Particle contamination primarily affects the lubricant's ability to control friction and wear. Obstructing the separation of moving components and interfering with chemical oil films provided by anti-wear and EP additives, particle contamination can substantially increase the rate of abrasion, adhesion, and surface fatigue. The effects of particle contamination are slow and imperceptible.

In fact, the loss of the lubricant's functional ability to reduce wear and friction is often overlooked due to the slowness with which particle contamination affects the system [22]. The best safeguard against contaminants is a clean, dry circuit breaker operating environment. If operating circumstances do not permit this environment, select sealed enclosures or shields to keep contaminants out. If humidity is a problem, consider selecting a lubricant with a good rust inhibitor. Harsh environments can sometimes be overcome by increasing the frequency of lubrication maintenance. More information on various sources of oil contamination is presented in Section 3.3.

2.2.4 Lubricant Incompatibility

Incompatibility occurs when a mixture of two greases shows physical or performance characteristics significantly inferior to those of either grease before mixing. Additional information is given in Addendum 2: Grease Compatibility. It includes the tables of compatibility of grease-based oils and thickeners.

An incompatibility issue could be reflected in reduced lubricating performance due to the modified composition of the fluids and additives from intermixing. Usually, problems are not obvious until the lubricated apparatus is in use. However, if two incompatible greases are mixed, lubrication failure is inevitable. Changes in physical properties would be reflected by a softening or hardening of grease, decreased shear stability upon mechanical working, and even reduction in thermal stability identified by a reduction in dropping point. Soft or runny grease is an indication of the problem. A grease that is much thicker than its original consistency may indicate incompatible mixing [6].

Grease color should also be observed. For example, if the original lubricant was green but has turned brown by the time of inspection, incompatible greases may have been mixed. Incompatible greases are a factor in many failures. Polyurea and lithium-based greases, for example, break down rapidly when mixed. It is best to know in advance which types of greases can be used together and which should not. Some greases cannot be mixed with others even when both types meet specifications. Unless this incompatibility is understood and accounted for, a switch to different grease can have serious consequences.

2.2.5 Preventative Measures

Storage conditions have the strongest influence on the rate at which the lubricant degrades. The storage environment strongly affects the shelf life of lubricants and greases [23]. The characteristics of some greases may change in storage. Grease may bleed, change consistency, or pick up contaminants during storage. Storage conditions may accelerate lubricant deterioration. Using proper technique to store the lubricants is important to avoid a premature deterioration of the products used in the field. Additional information is given in Addendum 3: Storage and Shelf Life of the Lubricants.

The person selecting the grease should investigate which solid or liquid additives, such as the extreme pressure (EP) type, will do the job well. For example, to lubricate open gears, grease containing a tacky additive must be used to keep the grease in place. The grease should be compatible with other materials used with the parts being lubricated. Manufacturers may specify the type of grease to be used in their equipment.

A common misconception among maintenance personnel is that it is better to over-lubricate than to under-lubricate bearings and matching parts. Both methods are undesirable. Under-lubrication risks metal-to-metal contact. Overlubrication pushes excessive grease into the cavity that causes heat buildup and friction as the moving elements continuously try to push extra grease out of the way. To ensure that moving parts are not over- or under-lubricated, follow the manufacturer's instructions, which applies both to grease and oil lubrication. In electrical applications, over-greasing may lead to contamination of nonconductive parts with grease, which will adversely affect dielectric properties of insulating materials.

REFERENCES

1. Salinas A. R., Desai B., Harley J., Kinner B, Lebow M. 2015. Circuit Breaker Mechanism Lubricant Performance Assesment: Investigation and Field Experience, 82nd International Conference of Doble Clients, 20. https://www.doble.com/wp-content/uploads/Circuit-Breaker-Mechanism-Lubricant-Performance.pdf
2. Bartlett D. 2012. Maintaining Low and Medium Voltage Circuit Breakers – Lubrication. https://avotraining.wordpress.com/2012/02/22/maintaining-low-and-medium-voltage-circuit-breakers-lubrication/
3. Demaria T. 2007. Circuit Breaker Lubrication in the Field, *NETA World Magazine*, Summer issue: 7–10, https://www.doble.com/wp.../Circuit-Breaker-Mechanism-Lubricant-Performance.pdf
4. Lubrication Guide, 2003. The Timken Company. http://www.timken.com/industries/torrington/catalog/pdf/general/form640.pdf
5. Ball and roller bearings, 2002. *Lubrication*, Publications of NTN Corporation, Japan, Chapter 11: A72–A79.
6. *Lubrication of Power Plant Equipment*, Facilities Instructions, Standards, and Techniques (FIST). 1991. Vol. 2–4, United States Department of the Interior Bureau of Reclamation (USBR), Springfield, VA: 33. http://www.usbr.gov/power/data/fist/fist2_4/vol2–4.pdf
7. Holzhauer R. Exclusive Guide to Synthetic Lubricants, *Plant Engineering Magazine*. http://www.plantengineering.com/industry-news/top-stories/single-article/plant-engineering-magazine-s-exclusive-guide-to-synthetic-lubricants/1428b70360.html

8. Pirro D.M., Wessol A.A. 2001. *Lubrication Fundamentals – Second Edition, Revised and Expanded*, Taylor & Francis Group, LLC.
9. Castaños B., Bazurto C., Peña-Parás L., Maldonado-Cortés D., Rodríguez-Salinas J. 2018. Characterization of Tribological Properties of Greases for Industrial Circuit Breakers, *Tribology in Industry*, Vol. 39, No. 4: 559–565.
10. Timken Bearing Damage Analysis with Lubrication Reference Guide. 2015. The Timken Company. https://www.timken.com/pdf/5892_Bearing%20Damage%20Analysis%20Brochure.pdf11. Lubricant Failure=Bearing Failure, 2009. *Machinery Lubrication* magazine, January/February. http://www.applied.com/site.cfm/Lubricant_Failure=Bearing_Failure.cfm
11. Lubricant Failure=Bearing Failure, 2009. Machinery Lubrication magazine, January/ February. http://www.applied.com/site.cfm/Lubricant_Failure=Bearing_ Failure.cfm
12. Bearing Failure Due to Over Lubrication. http://www.belray.com/bearing-failure-due-over-lubrication
13. Fitch J. 2011. Grease Dry-out: Causes, Effects and Remedies, *Machinery Lubrication*, August issue. https://www.machinerylubrication.com/Read/28517/grease-dry-out-causes
14. Bearing failure: causes and cures, Barden Precision Bearings. http://www.schaeffler.com/remotemedien/media/_shared_media/08_media_library/01_publications/barden/brochure_2/downloads_24/barden_bearing_failures_us_en.pdf
15. Ball & Roller Bearings: Failures, Causes and Countermeasures, JTEKT Corporation. www.koyousa.com/brochures/pdfs/catb3001e.pdf
16. Weigand M. Lubrication of Rolling Bearings – Technical Solutions for Critical Running Conditions. http://www.machinerylubrication.com/Read/844/lubrication-rolling-bearings
17. Inadequate grease lubrication in bearings: water contamination and debris contamination. 2009. Timken Automotive TechTips, Vol. 3, Issue 5, Part 3 of a 3-Part Series. http://www.timken.com/en-us/solutions/automotive/aftermarket/lightduty/TechTips/Documents/Vol3Iss5_Inadaquate_Grease_Lubrication_Part3of3.pdf
18. Booser, E., Khonsari, M. 2013. Grease and grease life. In *Encyclopedia of Tribology*; Springer, Heidelberg, Germany: 1555–1561.
19. Nailen R.L. 2004. Lubrication: Important for circuit breakers, too, *Electrical Apparatus*, March issue. http://www.findarticles.com/p/articles/mi_qa3726/is_200403/ai_n9361544
20. Neale M.J. 1995. *Lubrication: A Tribology Handbook,* Society of Automotive Engineers, Butterworth-Heinemann Ltd, Oxford, England: 640.
21. Johnson M., Spurlock M. 2010. Strategic oil analysis: Estimating remaining lubricant life, Tribology and Lubrication Technology, January: 22–30. http://www.lubetechnologies.com/assets/article—stle–best–practice–managing-lubricant-degradation.pdf
22. Bots S. 2013. Grease Analysis: Early Warning System for Failures and Proactive Maintenance Tool, *Machinery Lubrication*, No 2, February issue. http://www.machinerylubrication.com/Read/29284/grease-analysis-system
23. Snyder D.R. 2006. Unearthing root causes, *MRO Today Magazine*, December/January issue. http://www.progressivedistributor.com/mro/archives/Uptime/UnearthingCausesD05J06.htm

Lubrication of Mechanical Components in Wind Turbines

3

Wind turbines represent a challenging application for lubrication for multiple reasons. One reason is the serious difficulty of providing timely maintenance. The wind farms are usually located in remote areas; the turbines are very tall and maintenance would require special equipment to reach the parts requiring maintenance.

The ambient conditions in which wind turbines operate and the loads and vibrations to which components are exposed are relatively harsh. These conditions represent serious reliability challenges for an installed wind turbine fleet. Harsh environmental conditions include contamination from dust and wear debris, wide temperature range, high humidity and water ingress, and corrosion caused by offshore environments.

Designs of the new generation of multi-megawatt wind turbines require larger blade diameters and tower heights, increased pitch control, and a greater focus on offshore locations [1], which presents added complications and new challenges to bearing and lubricant manufacturers for wind turbines.

Multiple factors should be considered and analyzed to properly select the greases for wind turbine bearing lubrication. Among these factors is an understanding of the bearings' design and typical failure modes, lowest and highest ambient operating temperatures, servicing frequency, and method of application. A very serious factor is to ensure compatibility of the lubricants used during service with those applied at the factories.

Tribological challenges for wind turbines include complex failure modes involving bearings, gears, and lubricants inside wind turbine gearboxes. The following variables should be considered [2]:

a. wind speeds and directions;
b. intermittent operation with many starts/stops;

    c. high transient loads from wind gusts, grid engagement, and braking;

    d. high torque and low-speed input;

    e. remote locations that create maintenance difficulty.

The lubricants have to meet the necessary requirements from different subsystems such as main bearings, generator bearings, pitch bearings, yaw bearings, and yaw gears in diverse operating conditions. Bearing greases, for example, should be easy to pump and allow precise metering in centralized lubricating systems, thereby attaining good grease distribution.

There are currently two approaches to the development of greases for main bearing, blade pitching, yaw bearings, and generator bearings: developing a multipurpose grease suitable for the lubrication of all components, or developing a number of greases designed for each given component. Each approach has its pros and cons. The balance between ultimate bearing reliability and ease of use and ordering for service technicians has to be considered carefully [1].

When choosing a lubricant suitable for a particular wind power plant, the operator has to consider several important factors. Good wear protection even under vibration increases lifetime of the bearings during periods of idleness. Also, when the power station runs at low speed, wear is induced due to the lack of a sufficient hydrodynamic lubricant film. A good lubricant must contain suitable additives to neutralize these effects. Finally, it has to be ensured that the lubricant is compatible with the elastomers involved and covers the wide service temperature range of −40°C to 150°C (−40°F to 300°F) [3].

3.1 LUBRICATION OF WIND TURBINES

In a wind turbine there are multiple points that require lubrication, and depending on location and function of the parts there are different types of lubricants that should be used at every specific location. They should be chosen together with coatings if any are applied. Among mechanical parts to lubricate in wind turbines are bearings and threaded connections. Other parts, such as brakes, shrink discs, and various service parts also require protection from corrosion and lubrication.

3.1.1 Lubrication of Wind Turbine Bearings

In wind turbines bearings are everywhere: blade bearings, pitch bearings, yaw bearings, main shaft bearings, generator bearings, etc. Bearings are in high-speed/low-speed gearboxes [4]. Lubrication points for wind turbine for grease lubrication of different bearings are presented in many sources [5,6]. There are multiple issues to consider when choosing proper products and techniques for lubrication of wind turbine bearings.

Different Operating Conditions. The essential bearings of a wind power plant operate under very different operating conditions and therefore pose very different requirements regarding lubrication [3]. The main bearing rotates slowly but is subject to high loads and vibration. The generator bearing, by contrast, needs to cope with high speeds and high temperatures. Pitch and yaw bearings are subject to high loads as well, but they also perform oscillating motion under strong vibration. Due to these varying requirements, wind power plant operators have often had to resort to a variety of greases up to this point.

Qualities of the Lubricants. The lubricants in bearings application should provide high load carrying capacity, low friction, and protection against wear, fretting corrosion, groove formation (Brinell effect), and moisture. Lubricants should perform properly at low temperatures (up to −40°C) in cold climates and protect from offshore corrosion. Conditions and requirements for the wind turbine bearing greases are summarized in Table 3.1 [6]. Special bearing greases provide long-term lubrication of the bearing, wear protection in mixed friction due to solid lubricants and EP additives, as well as corrosion protection. An important issue for bearings lubrication is high loads occurring due to the wind gusts and inner vibrations in the turbine. The intermediate bodies of the bearings, like balls or rollers, are pressed against the outer and inner rings under vibrations so that the grease is squeezed out of the highly loaded contact zones. This generates wear marks in the form of the contact area of the intermediate bodies against the rings, which damages the bearings severely [7,8].

Lubrication Recommendations. Most wind turbine parks use turbines from more than a single manufacturer, so different lubricant recommendations have to be taken into consideration. And most manufacturers offer various turbine models, which are often used in parallel. For the operator this means that he has to spend a lot of resources on logistics, storage, and grease disposal, plus an increased risk of lubricants being mixed up. Most turbines are still lubricated manually, so service technicians have to carry a variety of lubricants. Additional information useful for the operators and service technicians is included in Addendums 1–3.

TABLE 3.1 Bearing lubrication challenges

	BEARING LOCATION			
CONDITION	PITCH/YAW	MAIN	GENERATOR	CHALLENGE
Wide temperature range, °C (°F)	−40 to 50 (−40 to 122)	−40° to 70° (−40 to 158)	−40° to 100° (−40 to 212)	Lubricant viscosity
Load capacity	X	X		Composition
Slow speed	X	X		Oscillation
Shock loads	X	X		Working T Range
Moderate/High speed			X	
Vibration	X	X		Fretting
Salt water, sand	X			Corrosion, abrasion
Humidity/ Condensation		X	X	Corrosion
Long life		X	X	Wear

Source: Adapted from H.R. Braun. 2011. Wind Turbine Grease Lubrication, NREL Wind Turbine Tribology Seminar, Broomfield, Colorado, November 15–17:1–45. http://www.nrel.gov/wind/pdfs/day1_sessionii_3_exxonmobil_braun.pdf

3.1.2 Lubrication of Threaded Connections

Safe, reliable wind turbines depend on strong threaded connections. Threaded connections are found practically everywhere in wind turbines, in nacelle frame joint, in front and rear main journal bearings, in pitch bearing fasten bolts, and in gearbox, drives, and cooler units. Yaw bearing fasten bolts, tower segment flange bolts, foundation anchor bolts all have threaded connections. Frequent problems with threaded connections often result in failures, such as tightening difficulties and thread damage, inconsistent or loose clamping force, broken bolts and base parts, such as flanges or plates, as well as difficult disassembly and destroyed threads.

Antifriction coatings (AFCs) and lubricants should be used on threaded connections on pitch and yaw bearings, tower, and other parts. These materials are thread pastes and anti-friction coatings (AFCs). Such materials provide consistent tightening torque and friction coefficients, high load-carrying capacity, anti-seizure and corrosion protection, and wide temperature range. Lubrication products for this particular application enable precise conversion of tightening torque into force during assembly and non-destructive dismantling of connections even after long use under high temperature, as well as equipment corrosion protection [9].

Using good lubricants for bolts and fasteners can help solve problems. They will increase reliability by reducing the effects of root causes, including: seizure and abrasive wear, uneven coefficient of friction, fretting corrosion, embrittlement failures of substrates due to use of low-melting-point metals, such as lead, tin, and copper, and stress corrosion cracking. Some of these materials are Molykote 1000 and Molykote G-Rapid Plus pastes from Dow Corning [10].

3.1.3 Lubrication and Corrosion Protection of Brakes, Shrink Discs, and Service Parts

For brakes and shrink discs, various pastes and AFCs provide corrosion protection, friction reduction, noise reduction caused by stick slip, and wear protection. Clean and dry AFCs provide an excellent combination of corrosion protection and lubrication. These coatings are easy to apply and they provide strong adhesion. The pastes suppress stick slip, prevent seizure and scoring, and reduce fretting corrosion. For parts difficult to reach and for added service safety, dry-film lubrication, such as antifriction coatings, is a suggested product.

3.1.4 Lubrication of Wind Turbine Blades

The gearbox is not the only part of the turbine that requires lubrication. The generator bearings and blade bearings also require lubrication, and there are lubrication points on the blades. Wind tower blades have bearings that will essentially allow operators to optimize the blade angle to match wind speed. The main shaft bearing and yaw and pitch drives also require lubrication. The turbines also use a hydraulic system that provides a braking mechanism for a unit, but can also be used for hydraulic pitch control on the blades [11].

As with many other bearings, the blade bearings are lubricated with an automatic greasing system. Greases for blade bearings and azimuth bearings have to show the following properties: high load resistance, resistance to vibrations as a result of the lightweight design and unsteady wind conditions, medium to low speeds, high temperature resistance and broad service temperature range, resistance to corrosive environments, good compatibility with the sealing materials and varnishes involved, and good pumpability in centralized lubricating systems [12].

3.1.5 Environmental Factors in Lubricant Selection

The wind turbines could operate almost 24/7 in extreme environmental conditions, such as in the subzero temperatures of the far north or the scorching heat of deserts. Installed in deserts the blades and mechanisms of wind turbines are continuously pummeled by sand. Wind turbines can be exposed to high humidity, with saltwater seeping into the tower. In such extreme environments the choice of lubricants is a very important issue in providing a long servicing life of the equipment.

The gearboxes are designed to last 20 years. However, since they endure great environmental assault, gearboxes are beginning to fail within seven years in service. Lubricant degradation seems to play a major role in these failures, though the reasons for the early failures are not always fully understood. Several environmental factors may adversely affect the performance of various lubricants used in wind turbines whether it is main gearbox oil, hydraulic fluid, or grease.

3.1.5.1 Wet and corrosive environment

Wind farms are commonly located in coastal areas onshore and offshore surrounded by sea water, which results in an increased potential for saltwater ingress. Therefore greases for wind turbine lubrication must provide excellent rust and corrosion resistance even in a saltwater corrosion environment.

A lubricant for use for wind turbine lubrication in coastal areas should pass an industry standard test ASTM D6138, which is SKF EMCOR test "Corrosion-Preventive Properties of Lubricating Greases Under Dynamic Wet Conditions." This test is designed to determine the anti-corrosion properties of greases when exposed to waters of varying quality [13]. ASTM D6138 is conducted in contact with double-row self-aligning ball bearings. At the end of the test, the bearing raceways are examined and the degree of corrosion is rated against a defined rating scale of 1–5, with 0 being no corrosion and 5 representing heavy corrosion with corroded areas covering more than 10% of the running track surface. Advance corrosion-preventive greases would provide a rating of 1 or less than 1 with synthetic seawater.

3.1.5.2 Thermal effect

Wind turbines work in climate ranges from extreme arctic conditions to the high ambient temperature of deserts. Temperatures from <–40°C to >50°C are not uncommon, often with wide seasonal and daily swings. One of the most critical requirements for the wind turbine lubricant is the ability of the lubricant to function over a wide operating temperature range.

Most open gears of wind power plants are still lubricated by hand. However, maintenance can be reduced to keep downtime to a minimum. Today's typical wind turbines are fitted with centralized greasing systems of progressive or single-line type with narrow diameter feeder lines which distribute the grease to the lubrication points. Therefore it is important for the lubricants to remain fluid and have excellent low-temperature pumpability even at lower ambient temperatures and improved wear protection at higher operating temperatures [14].

> *High Temperatures.* If the lubricants used for lubrication of different parts of a wind turbine are exposed to high temperatures for extended periods of time, the damaging consequences are unavoidable. High oil temperatures reduce the lifetime of oil. It is a known fact that if the maximum rated oil temperature is constantly exceeded by 10°C, oil lifetime is reduced by half. Therefore, mineral lubricants cannot be used if oil temperature may exceed 80°C. For hot climates, an oil cooler for the lubrication system may be needed.
>
> *Cold Temperatures.* In the last decade it was found that the Arctic is a very attractive area with massive potential to generate clean, renewable energy. Due to the high average wind speed (around 30 km/h) and the fact that colder air is denser, arctic wind carries more kinetic energy. But the environment is very hostile: temperatures can drop to −25°C (−13°F) and winds speed can be over 180 km/h. The wind and cold make the turbines wear faster than in other locations. Designing a wind farm to handle these conditions is challenging [15].

A very important issue to the operation of wind turbines in cold weather is the behavior of the lubricating oil. Gearboxes, hydraulic couplers, and dampers suffer from long exposure to cold weather. As the temperature goes down, the viscosity of the lubricants and hydraulic fluids increases up to a point where at −40°C (−40°F), a chunk of heavy gear oil could be used to pound nails [16].

Damage to gears will occur in the very first seconds of operation where oil is very thick and cannot freely circulate. In addition, due to an increase in internal friction, the power transmission capacity of the gearbox is reduced when the oil viscosity has not reached an acceptable level [17].

Standard hydraulic oils become highly viscous at low temperatures. Due to the high viscosity of standard oils in low temperatures or to different properties of cold-temperature oils, turbine start-up may be delayed at higher wind speeds which will impact overall turbine performance. Using the wrong lubricants and greases has damaged bearings and gearboxes during low-temperature operation [18].

In cold climates it is recommended to use synthetic lubricants that are rated for cold temperatures. All wind turbine manufactures recommend specific lubricants based on their particular turbine design. The wind turbine operator is encouraged to obtain specific certifications prior to their use of the lubricants, though in most cases these lubricants have been tested. To avoid cold-related problems, in modern wind turbines surface-heated gearboxes and gearboxes with immersed heaters with constant oil circulation, generator heaters, and heaters for the cabins containing control electronics are used [19].

The latest developments present new wind turbine specialty lubricants for gear rim/pinion drives on pitch and yaw bearings that can offer good pumpability and metering in central lubrication systems to temperatures as low as −30°C (−22°F), thus contributing to the increased reliability of turbines at low temperatures [20].

3.1.5.3 Desert climate

Desert climate regions include the Sahara, Arabian, Syrian, and Kalahari deserts, large parts of Iran, northwest India, the southwestern United States, northern Mexico, the Kyzyl Kum and Taklamakan deserts of Central Asia, and much of Australia, where wind turbines are installed.

Because of the environment, additional risks of wind turbine failures arise in deserts. Dust, sand, and temperature extremes cause an elevated percent of failures and an increase in repair parts required to maintain wind turbines. Parts subject to friction fail with greater frequency in the desert than under onshore and offshore conditions. In a desert, a combination of heat, dryness, and fine particulate matter causes failure of various wind turbine parts and systems. Lubricants and lubrication systems are also exposed to damaging desert conditions. Ambient air that appears clean is actually laden with fine dust.

Extending the service life of the oil in the desert requires planned servicing of oil filters. Replacement of gearbox lubrication filter and oil changing operations must be done more frequently than what is recommended for non-desert conditions. Additionally, oil lubrication and cooling systems are less effective, and oil evaporates at the high temperatures generated during turbine operation. The greasing system of main bearings should not be lubricated unless it is sealed to avoid the attraction and accumulation of dust and dirt [21].

3.2 LUBRICATING OILS FOR WIND TURBINE APPLICATION

3.2.1 Basic Requirements for Gearbox Oil Qualities

The correctly selected oil can potentially extend the life of the wind turbine gearbox, reduce downtime, and lower maintenance costs. To maximize life of the gearbox, one of the most important components of a wind turbine, the gear oil should provide long-term gear and bearing wear protection. One of the most important gearbox oil qualities is *oxidation stability* to extend service life; it should provide rust and corrosion protection.

However, lubricating the gearbox of a wind turbine poses many challenges that have a direct impact on the required properties of the lubricant. Weight restrictions such as compact design and high load-handling capabilities require excellent protection against *micropitting* and *scuffing*. Such oils should have extreme pressure additives with a high load-carrying capacity and additives that prevent micropitting.

Extended oil drain interval demands make lubricant *property retention* important as well as long-term protection against *ageing*. Fine filtration systems with mesh size of 10 μm or smaller make oil *filterability* under dry and wet conditions critical.

Additional qualities are required for off-shore applications, such as outstanding rust and corrosion protection from salt water. Since wind turbines operate in wide range of environments—extreme temperatures and intermittent operations—the lubricants should provide *flowability* at low temperatures and *anti-wear capability* at high temperature.

The minimum requirements for a good wind turbine gear oil include long oil life (minimum 3–5 years), constant wear performance as the oil ages, thermal and oxidation stability, and resistance to sludge formation [22]. These oils should perform in a wide temperature range, providing cold startup and high operating temperatures.

Good oil should display *stability* with water and condensation, providing rust and corrosion protection, and oil should not affect the filter service interval. *Cleanliness* of new and used oils should be no less than the limits established in national and international standards. See additional information on the oil cleanliness standards in Section 3.3.3.

Wind turbine oil should be compatible with elastomers and paints, passing static, dynamic, and long-duration tests (at least 1,000 hrs), foam tests, micropitting tests, and bearing anti-friction tests. The oils should be tested for formation of residues under the influence of water and temperature.

3.2.2 Wind Turbine Gearbox Oils Meeting Requirements

Mineral oils cannot meet the basic demands for use in wind turbine gearboxes. *Synthetic* oils are preferable since they offer wide range of benefits, including improved thermal resistance, better viscosity characteristics, product longevity, and longer machine component life. To formulate gear oils, different base oils such as polyalphaolefin (PAO), polyglycol, or rapidly biodegradable ester are used. PAOs provide excellent viscosity index and low pour point. These properties make them a fluid of choice for applications characterized by wide ranges of operating temperatures. There have been hydrolysis issues (breakdown in the presence of water) with PAO/ester blends, making selection of hydrolytically stable products a critical issue. PAGs (polyalkalene glycol) offer improved resistance to micropitting but have compatibility problems with coatings and seal material [23].

However, many of these traditional synthetic oils cannot meet the new requirements created by changes in the wind power industry. Wind turbine operators are using new, higher performance synthetic oils with increasing frequency. To meet and even exceed the new standards the lubricant industry is developing specialty products. The new lubricants offer high thermal resistance and resistance to oxidation; lower change in viscosity at rising or falling temperature; lower friction coefficients; high wear protection for bearings and gears; good load-carrying capacity in bearings and gears, and low residue formation. These lubricants should offer extended service life and economical operation.

Developing these lubricants requires knowledge of base oils, additives chemistry, and which combination of additives to use. The more pure the molecular structure of the base oil, the better the lubricant [24]. Compared to standard gear oils, modern gear oils can reduce temperatures by as much as 20°C (68°F) and power losses by as much as 18%. Lowering the friction component in a wind station improves efficiency and increases power output.

New high-performance synthetic oils are subject to the tests of original equipment manufacturers (OEMs) and must meet requirements of a number of universal standard tests on viscosity, pour point, foaming characteristics, and steel and copper corrosion. The elevated requirements include multiple additional tests that the oil must pass prior to choosing a particular oil to use in gears [22,25]. The list of some ASTM tests qualifying oil for wind turbine application are presented in Table 3.2.

Most of the wind turbine gearbox manufacturers have compiled or are in the process of compiling new lubrication specifications. These specifications are more stringent than those for industrial gear applications and more accurately reflect true operating conditions, including low-temperature conditions. Performance expectations for lubricants used in offshore wind turbines are higher due to demand for extended life [23].

An additional test is the German Standard DIN 51 517, Part 3 [26], which defines the requirements for gear oils that are exposed to high loads. This standard specifies a scuffing load stage greater than 12 for gear oils. Because gear oils should also be suitable for lubricating the rolling bearings in the gearbox, this standard also contains rolling bearing test rig FE 8, developed by the rolling bearing manufacturer FAG. The FAG FE 8 test rig can be used to assess the anti-wear properties of an oil and its effect on the rolling bearing service life. In this test, the wear of the rolling elements should not exceed 30 mg.

The assessment of gear oil performance for wind turbines also includes tests developed by Gear Research Center (FZG). These tests measure scuffing load resistance (DIN 51534, FZG Scuffing Test) and micro-pitting resistance (FVA Proc No. 54, FZG Micropitting Test). The tests determine anti-wear properties of the lubricant at low gear speeds, as the planetary gear stage is run at the lowest speed and better-performing lubricants fall within the low-wear category. Gear efficiency is determined to a large extent by the friction characteristics of the lubricating oil. The friction coefficients of different base oils can be seen in the result of the FZG test rig [27].

3.2.3 Automatic Greasing Systems for Wind Turbines

Wind energy systems need proper lubrication to function optimally. Vibration, high mechanical loads, contamination, and moisture are all threats to bearing and gear service life. Wind turbines can reach more than 100 meters off the ground and are often in remote and difficult-to-access locations. Wind turbines can be challenging and expensive to service. The solution is an automatic greasing system, which is used to lubricate many motor and generator bearings in wind turbines.

Automatic lubrication systems promote improved operating times and longer maintenance intervals, typically one year or longer. Additional benefits are lower costs for repairs, spare parts, and lubricant, and greater bearing

TABLE 3.2 ASTM standards/tests qualifying oil for wind turbine application

NO.	STANDARDS	TEST NAME	TESTED PROPERTY	TEST CONDITION	TEST PASSED
1	ASTM D3336—05e1	Test Method for Life of Lubricating Greases in Ball Bearings at Elevated Temperatures	Lubricant Life	Number of hours until failure of grease at temperature up to 371°C @ 10,000 rpm.	Product life estimated as number of hours before 10% of grease failed or hours to 50% failure
2	ASTM D2893, ASTM D5704, ASTM D2270	Oxidation Test, L-60-1 Thermal Oxidation Test, Viscosity Index	Oxidation Control, Stability, Viscosity Index	Hours required for % viscosity and TAN increase	40–120 hours
3	ASTM D6821 ASTM D2983 ASTM D97	Low T viscosity Pour point	Low Temperature Flow	Viscosity measured at temperature from −40°C to −26°C	
4	ISO 13357	Filterability of Lubricating Oils	Filterability	Oil with 1% water preheat to 70°C (158°F) pass through a 3 micron filter	Pass determined by pressure differential, oil unacceptable if oil blocks filter
5	ASTM D1401 ASTM D2711 ASTM D6422	Demulsibility of industrial gear oils Water tolerance	Water Tolerance	Water added to oil at 130°F, stir 5 min at 1500 rpm, time to water/oil separation in min Residue after contact with 1% water	15 min No residue in 24 hours @ 71°C (160°F)
6	ASTM D-892	Foam Tendency Test	Foam Control	Measure volumeric % of foam formed in testing cell filled with oil and turned gears for a certain time at 25°C	Pass if < Vol.15% of foam
7	ASTM D3427	Air Release Test	Air Release		<20 min
8	ASTM D7038—15	Moisture Corrosion Resistance of Automotive Gear Lubricants	Rust & Corrosion Protection	Signs of rusting in roller bearings rotating for 110 hours in oil in the presence of water (distilled, salt, or synthetic sea water)	On rating scale from 0 (no rust) to 5 (heavy rusting), 5 is not acceptable

life from regular, exact amounts of lubrication. An automatic greasing system minimizes or eliminates safety risks associated with hard-to-reach lubrication points, increases corrosion protection from the elements, and significantly reduces the amount of wasted lubrication. A special gearbox oil filter, separate from the normal oil cooling system, ensures high oil cleanliness. This is a key factor in desert or arid conditions where airborne dust can get into gearboxes, act as an abrasive, and eventually lead to contact fatigue failures.

Automatic lubrication systems provide lubricant supply more reliably and precisely to moving components in the nacelle when compared to manual lubrication. By delivering the smallest effective amount of lubricant reliably to all friction points while the machine is running, automatic lubrication systems reduce friction inside bearings and help prevent contamination. The result is optimized bearing service life over the long term, more turbine uptime, and reduced manpower costs, which helps to make wind farms more profitable.

Automatic lubrication systems can offer a quick return on investment by increasing turbine system availability, extending maintenance intervals and preventing failures of major components [28]. Additional savings can be achieved through proper lubricant handling and consumption. Automatic lubrication systems built by SKF and Lincoln are popular in the wind energy industry.

As maintenance normally cannot be carried out without taking the turbine offline, there are only two options to lubricate the gears of the rotor blade: (1) lubrication during the natural idle phases (no wind); and (2) compulsory maintenance intervals, associated with additional losses of production.

The idea of adapting the lubrication task to natural idle phases is attractive to wind farm operators but not at all to designers or maintenance staff. This is because production phases may be extremely long during windy periods. These periods tend to be even longer for offshore plants than for onshore installations. In such cases, the lubrication film could degenerate and fail before the end of the natural wind production phase, thus causing damage to the gear drive.

A special lubrication system, MiCRoLuBGeaR [29] has been designed to enable the independent re-lubrication of different teeth in open gears and to be used in wind turbine's pitch gears. It avoids energy generation losses and lubricates the pitch at the zero-degree position. MiCRoLuBGeaR, combined with the correct automatic lubrication system, lubricates the tooth in contact while the wind turbine operates and the pitch system is working.

SKF's condition-based lubrication provides remote and automatic lubrication of the hard-to-access bearing systems of a wind turbine. This is an interface enabling a connection between the SKF condition monitoring system (SKF WindCon) and SKF or Lincoln lubrication systems (SKF Windlub [30]). Reacting to the problems detected by the condition monitoring system (CMS),

SKF's condition-based lubrication allows a lubrication pump to initiate additional lubrication cycles on top of the existing time-based cycle.

A condition-monitoring specialist can set the proper alarm settings to trigger additional lubrication cycles. It allows the SKF WindCon system to monitor lubrication pumps and components, including pump status and grease levels. Operators are notified immediately if failures such as empty or blocked pumps or torn feed lines are detected. This early detection of lubrication failures, working in combination with automatic lubrication, will naturally avoid unnecessary and inconvenient maintenance operations [31].

3.2.4 Biodegradable Lubricants for Use in Wind Turbines

In recent years interest has grown for biodegradable and non-toxic lubricants, which were initially developed to avoid the polluting effects on rivers, lakes, and aquifers for applications with total loss or risk of leakage. Currently many leading companies manufacturing oils and greases are developing biodegradable lubricants for use in wind turbine applications.

The use of bio-lubricants in wind turbine gear may also offer improvements from a performance standpoint, especially under severe loading conditions [32]. In this study the tribological effects of using bio-oil base Biotelex with halogen-free and halogenated ionic liquids as additives were tested for use in wind turbines and compared against each other as well as against a fully formulated, high-performance synthetic oil. The study showed that the use of ionic liquids as lubricant additives resulted in significantly reduced friction and wear under the most severe conditions tested. The studies show that bio-lubricants can reach a technical performance similar to mineral-based lubricants in many applications, though there are some limitations related to extreme temperatures.

RSC Bio Solutions, a developer of environmentally acceptable lubricants and cleaners, recently announced the expansion of its EnviroLogic 800 series of biodegradable greases. The EnviroLogic 800 series are VGP-compliant, biodegradable, lithium complex greases designed to operate in severe outdoor environments and withstand corrosion, such as in wind turbines onshore and off. These greases demonstrate excellent low-temperature pumpability, EP and anti-wear protection, and oil separation stability during storage. They are biodegradable and will not cause harm to aquatic life or animals [33].

Spanish companies Verkol, S.A. and Sotavento Galicia, S.A. have been working in the field of biodegradable, non-toxic, and renewable lubricants. Currently they are at the testing and demonstration phase of the operation where field tests are being developed at the Sotavento Experimental Wind Farm with an oil for gearboxes and a grease for slewing rings. The biolubricants are based on high oleic sunflower oil (HOSO 83%) [34].

3.2.5 Lubrication Maintenance

In wind energy industry, routine maintenance and relubrication is demanding and specialized work. The gearbox is situated just where the winds are the strongest—as high as 100 meters (~300 feet) from the ground or sea level. In addition, offshore installations encounter rough seas. The maintenance engineer will have to gain access up the tower via an internal ladder (or elevator in some cases). Oil drain intervals are usually between 8 and 12 months. It is expected that new-generation oils for offshore applications could have a drain interval of up to three years. With the high growth rate of wind power, understanding the impact of proper lubrication on high-stress equipment is essential to achieve a high availability.

3.2.5.1 Maintenance procedures and time frames

Lubricating materials, procedures, and test methods are continually improving, providing invaluable tools for maintenance personnel that improve the operation and safety of wind power generators. To maintain reliable systems, a detailed inspection, testing, and sampling route for the gearbox main shaft bearings and yaw/pitch gears should be developed for uptower inspection processes. Typical time frames for the inspection process are: monthly, quarterly, every 4–6 months, and annually. Recommended lubrication maintenance procedures are shown in Table 3.3. If oil contamination is an ongoing issue, routine weekly lubricating fluids sample cycles can help predict failure, allowing for a pre-planned outage window for component repair/replacement.

When maintaining collector substation and circuit breakers, electrical maintenance programs that employ a lubrication component as part of the program should be developed. The program should be based on national consensus standards. Adopting a complete maintenance and lubrication plan will significantly reduce the risk of catastrophic equipment failure [35].

3.2.5.2 Education and skill standards of maintenance personnel

Another important issue is to educate and train technical personnel according to the standards developed for wind turbine technicians [36,37] to perform the manifold tasks of wind turbine maintenance including

TABLE 3.3 Recommended lubrication maintenance procedures for wind turbines

EVERY MONTH	EVERY QUARTER	EVERY 4–6 MONTHS	ANNUALLY
Inspect visually the machine's anchorage, alignment, oil levels, filters, pumps, valves, coolers, heaters, manifolds, and piping to connect components. *Sample and analyze vibration* on the rolling elements.	*Monitor the condition of the lubricating fluids*: sample and test gearbox oils and generator greases to determine wear and contamination levels for metals, viscosity, particle count, water/ moisture, total acid number, and oxidization/ nitration.	*Change* lubricating oil filters. *Apply* grease to gear teeth and slew bearings. *Check and top off* oil lubricants in yaw and pitch gears.	*Inspect contact and wear patterns* of gears and bearings on the rolling element. Use borescope and videoscope cameras.

lubrication. These standards include knowledge and understanding of lubrication (oil & greases), skills of greasing bearings and verifying fluid levels (gear oils and hydraulic fluids), replacing oil and breather filters, changing drying agents, and so on, according to the maintenance schedule. The standards summarize technical skills, knowledge, and abilities; all areas of expertise that workers must have to perform a given occupational task with excellence.

3.2.5.3 Post warranty maintenance lubrication

Most of the worldwide wind fleet is entering a post-warranty time period. The average age of wind turbines in North America will reach 7 years in service in 2020. This requires a well-planned preventative maintenance strategy, an essential part of which is thorough oil changes. In the harsh conditions in the wind industry, with extreme temperatures, heavy and varying loads, and exposure to water contamination, lubricants are the first line of defense. Oil changes should be scheduled in advance to be performed in warm weather, no colder than ~2°C (35°F), which will use less downtime and result in a cleaner oil change. It also would be better done in periods of low or no wind. Any preventative maintenance program should include used-oil analysis to identify turbine reliability issues such as lubricant quality and abnormal condition [38].

3.3 OIL CONTAMINATION IN WIND TURBINE GEARBOXES

There are many different ways for contamination to come into gearboxes of wind turbines. Contaminants can enter gearboxes during manufacturing, be internally generated, ingested through breathers and seals, and unintentionally added during maintenance. To limit the impact that contamination can have on components, all of these sources must be addressed [39,40].

The types of oil contamination are often defined by the location of the wind turbines. Desert wind turbines are exposed to airborne dust during the hot season and moisture during the rainy season. Offshore turbines are constantly exposed to moisture.

The level of gears and bearings damage that the particles in contaminated oil can cause depends on the particle's properties, such as size, hardness, friability, or ductility, which in turn depend on the composition of the particles. The higher the hardness of the particles (on a scale of 1 to 10, with 10 being the hardest), the larger the damage they produce. Environmental dust, which is mostly silica, has a hardness rating of two to eight. Quarry dust has a rating of five to nine. Common contaminants like rust and black iron oxides have a Mohs hardness rating of five to six. From the manufacturing process, tool steel has a hardness rating of six to seven. Silicon carbide and aluminum oxide have a hardness rating of nine.

3.3.1 Built-in Contamination

Contamination can enter gearboxes during manufacturing, Filters do not immediately remove built-in manufacturing debris. Consequently, permanent debris dents and other damage may occur during run-in, unless gearboxes are assembled in a clean room using clean assembly procedures and then filled with clean lubricant.

There are many sources of contamination to eliminate long before the gearbox is placed into service. For instance, the interior of gear housings should be painted with white epoxy sealer to provide a hard smooth surface that is easy to clean, seals porosity, and seals in debris like casting sand. All components for assembly should be properly stored in a dry area prior to assembly. All gears and bearings should be covered, and bearings should be stored on their sides.

All components should be cleaned prior to assembly. Initial cleaning should be done in an area separate from the clean room, followed by final cleaning in the clean room just prior to assembly. All components should be carefully inspected to ensure they are clean and rust-free before assembly. Special attention should be paid to bolt holes, oil passages, and other cavities that may contain dirt.

Gearboxes should be assembled in a clean room separate from any manufacturing processes such as machining, grinding, welding, or deburring. Windows and doors should be adequately sealed to prevent contamination ingression and the ventilation system should be filtered so it provides clean, draft-free air. The floor should be painted and sealed so it is easily cleaned and the overhead structure should be painted and dust-free. No tow motors should be allowed in the clean room because they invariably introduce contaminants.

3.3.2 Internally Generated Contamination

Contamination particles that are generated inside gearboxes are usually wear debris from gears, bearings, splines, or other components. The contamination results from various wear modes, such as micropitting, macropitting, adhesion, abrasion, or fretting corrosion.

Fretting corrosion may occur on gear teeth, splines, and bearings in wind turbines that are parked for extended periods. Wet clutches and brakes can also produce wear debris. To avoid gear oil contamination, wet clutches and brakes should have separate lubrication systems.

To minimize internally generated wear debris, it is recommended to use high-viscosity lubricants. The use of the following features in gearbox can also minimize generation of contamination particles: accurate and smooth surfaces and surface-hardened gears and splines. To prevent fretting corrosion the surfaces may be carburized or nitrided, which are heat treating processes that diffuse carbon or nitrogen into the surface to make it harder. External and internal spline teeth surfaces should be improved by being nitrided and force-lubricated. Annulus gears should be carburized or nitrided rather than through-hardened because through-hardened gears are relatively soft and prone to generating wear debris [39].

Many of the gearbox and generator manufacturers utilize epoxies and specialty glues to attach bearing covers, inspection ports, and filler caps. Many times trace to large amounts of these substances are found in grease and oil samples, and thus are implicated in bearing failures of gearboxes and generators. Large amounts of sand and epoxies contribute to excessive wear, overloading, and overheating that eventually leading to catastrophic bearing failure.

3.3.3 Required Oil Cleanliness for Wind Turbine Gearboxes

Cleanliness of gearbox oil plays a crucial role in life of the gearbox. For example, British researchers [40] showed that rolling element bearing life can be increased up to seven times by changing from a 40-mm filter to a 3-mm filter, thus providing better cleanliness of the oil. Their results also show that a gearbox must be clean after assembly or a fine filter will not be effective. Even before a filter can remove built-in contamination, gears and bearings can suffer permanent damage in as little as 30 minutes during run-in. There are many sources of contamination of the oil [39].

Lubrication systems should be properly designed and carefully maintained to ensure gears receive an adequate amount of cool, clean, and dry lubricant. Modern filters are compact and provide fine filtration and long life without creating large pressure drops. Offline filters provide fine filtration during operation and during turbine shutdown. Once the oil is clean, it should stay clean, provided the gearbox and lubrication system were properly designed and seals, breathers, and maintenance are adequate.

An industrial standard AGMA/AWEA 6006-A03 "Standard for Design and Specification of Gearboxes for Wind Turbines" [41] is written by the American Gear Manufacturers Association (AGMA) in cooperation with the American Wind Energy Association (AWEA). It provides guidelines for specifying, selecting, designing, manufacturing, procuring, operating, and maintaining gearboxes for use in wind turbines.

In Section 6—Lubrication, this standard discusses many elements of the lubrication system that determine oil cleanliness, and sets up the oil cleanliness levels in gearbox that comply with international ISO 4406:1999 Fluid Cleanliness Standard [42]. Three numbers defining oil cleanliness $a/b/c$ correspond to the number of particles larger than 4 micron (a), 6 micron (b), and 14 micron (c) respectively in particle content of 1 milliliter of oil. Required oil cleanliness for wind turbine gearboxes according to AGMA/AWEA 6006-A03 and ISO 4406:99 is: 16/14/11 for oil added to gearbox; 17/15/12 for oil from gearbox after factory testing and for oil from gearbox after 24 hour in service; and 18/16/13 for oil from gearbox in service.

To improve oil cleanliness in wind turbine gearboxes, it is recommended to use an offline filter system, which is simple and easy to install [43]. The benefits of the offline filter include improved lifetime of the gear oil and gearbox bearings, reduced wear and tear on gearbox bearings, and reduced risk of bearing damage due to poor oil cleanliness.

3.3.4 Water Contamination of Oil

Many experiments have shown water in oil promotes both micropitting and macropitting through loss of oil film. Water interferes with the pressure-viscosity coefficient of the oil, diminishing its ability to momentarily solidify in the contact area.

It was found that the primary cause reduced bearing life was occasional passage of microscopic water droplets under high pressure through the lubricating zone, resulting in local lubricating film breakdown. The number and the size of the microscopic droplets increase with the amount of water in lubricating oil, thus increasing the probability of water passing through the lubricating zone.

Excess water in wind turbine gear oil is associated with many negative effects [41], such as:

- Accelerated additive depletion
- Accelerated oxidation
- Interference with oil film formation
- Increased foaming
- Plugged filters
- Corrosion etch pits and initiate fatigue cracks
- Hydrogen embrittlement promoting fatigue cracks

Another result of water presence in oil, particularly with water concentrations above 300 ppm, is the tendency of oils to form residues at high temperatures in the form of sludge and varnish. This not only accelerates the aging of the base oils, it also causes additives to precipitate out or reduces their effectiveness. Thickened oil/water emulsions also suspend abrasive particles in lubricants and cause surface damage by indenting and scratching the metal surface, causing stress concentrations and disrupting the lubricant film.

Therefore it is important to avoid using oil which is susceptible to water contamination. Water presence in oil shortens the life of bearings significantly. It was found that ester-based lubricants and mineral oils with EP or AW additives are especially prone to absorbing water [44]. Properly sealing the gearbox can prevent water from entering into the gearbox.

3.3.5 Oil Filtration in Wind Turbines

There are three types of lubricants/oils in wind turbines: gear oil (in gear box), hydraulic oil (in pitch hydraulics), and lubrication oil (in main bearing). Multiple problems are most frequently caused by contaminants in the oil and can be avoided through offline fine filtration.

For transformers connected to the wind farms and used to feed electricity into the grid, an added issue could be also tap changer oil, when the contacts of the on-load tap changers are contaminated with deposits of oil degradation products. Oil maintenance may prevent downtimes. With a specially designed fine filter system it is possible to dry insulating oil while the transformer is running and, thus, to stabilize the dielectric strength and increase the operational reliability.

Advanced filter systems for wind turbines should fulfil the following requirements: high dirt holding capacity, absorption of water, and adsorption of varnish. Continuous high oil cleanliness can only be ensured by continuous offline fine filtration in conjunction with the inline filter. Only the offline principle allows a perfect gear oil flow rate/filter size relation. The oil flows through the filter body at an extremely slow pace so that even micro-fine particles settle down deeply within the filter insert [45].

Most gearbox manufacturers utilize a special gearbox oil filter that helps ensure maximum oil cleanliness. This is a key factor in environments where dust and debris can get into gearboxes and eventually lead to contact fatigue failures [2,35].

3.3.6 Gear Oil Contamination Control

There are several ways to control oil contamination and mitigate damage to oil-lubricated parts of wind turbine caused by contamination [46].

Desiccant Breather. Most external contamination ingression in wind-turbine gearboxes comes from the breather or vent port. Most OEMs require the use of a desiccant breather. These breathers include a particle removal element capable of eliminating silt-sized particles and a desiccating media often comprised of silica gel. This media is able to remove all traces of moisture from the air as it enters the gearbox. This type of breather provides bi-directional water removal by impregnating the silica gel with an indicator sensitive to the degree of moisture saturation. The change-out interval for a desiccant breather can be optimized by observing the color change within the desiccant.

Vented Breathers. Standard desiccant breathers are effective at preventing particle and moisture ingress. However, their life can be relatively short, depending on the operating environment. To improve this, desiccant-breather manufacturers have developed vented breathers, which are nominally sealed, preventing contamination ingress and preserving the life of the breather. Depending on application and environment, these so-called hybrid breathers can last as much as 5 to 10 times the life of a conventional desiccant breather.

Kidney-loop Filtration Systems. The next step in controlling gear contamination is permanently mounted dedicated kidney-loop filtration systems, which are vital to eliminate particle-induced failures from internally generated particles. Kidney-loop filtration systems remove particles and moisture in oil before returning it to the oil sump. Ideally, the filtration system should allow both particle and moisture removal, and have good particle capture efficiency to optimize the time between necessary filter changes the time between necessary filter changes.

3.4 WIND TURBINE BEARING AND GEARBOX FAILURES CAUSED BY LUBRICATION

Many factors may cause degradation of oil used in wind turbines to lubricate many bearings and gearboxes. The oil can become a source of failure of wind turbine components if it is not maintained properly. Oil degradation includes loss of additives, oil oxidation, and microbial growth among other factors. Additive molecules with great affinity to water form bonds with water molecules and contaminate the whole lubrication system. Oil oxidation is one of the most severe forms of oil degradation when oxygen atoms chemically react with oil molecules and produce acids and polymeric compounds as by-products. These acids promote corrosion of various parts of the wind turbine. Contamination of oil with microorganisms can quickly deplete oil quality and performance. Formation of atomic hydrogen from decomposition of lubricant molecules and additives may result in hydrogen embrittlement of steel parts.

3.4.1 Lubricant Roles in Triggering Bearing Failures

In wind turbines very complex failure modes involve bearings, gears, and lubricants inside wind turbine gearboxes. The failure may be caused by some component of the lubricant, by decomposition of grease or oil in service, by lack of lubrication in particular type of bearings, by contamination of the lubricant, etc.

Oil Oxidation. Oil oxidation is a series of chemical reactions both initiated and propagated by reactive chemicals (free-radicals) within the oil. If metal wear debris are present in oil, the rate of oil oxidation increases by two orders of magnitude; oil thickens and acidity increases, which promotes corrosion of various parts of the wind turbine. At first, the anti-oxidant additive package depletes and then the base oil oxidizes. The antioxidant additive is sacrificial—it is there to protect the base oil from oxidation. The most common antioxidant additives found in wind turbine gear oils are phenolic inhibitors, (these work to neutralize the free-radicals which cause oxidation) and aromatic amines (which work to trap free radicals) [47].

Hydrogen Embrittlement. Hydrogen embrittlement is considered as a potential failure mechanism involved with high strength steels. White structure flaking (WSF) or white etching crack (WEC) can bring about premature failure of roller bearings in wind turbine gear boxes. This type of flaking is induced by diffusing of hydrogen from the lubricant into the steel, causing the steel embrittlement. Investigations in REFs [48–50] confirm the formation of atomic hydrogen from decomposition of lubricant molecules, additives, and contaminants via tribochemical reactions in tribological contacts during surface-rubbing. Such a route of hydrogen uptake from lubricated surfaces has been considered as one of the root causes in promoting early failures of bearings and gears. It is reported that lubricant is decomposed by a chemical reaction with a fresh metal surface, which is formed by local metal-to-metal contact and thereby generates hydrogen.

Besides the well-known sources of hydrogen uptake, it has been also observed in lubricated sliding or rolling contacts, where a surface passive layer is continuously removed and the fresh steel surface remains exposed to hydrogen attack from lubricant and additives in the contact area.

Additives Depletion. Operating conditions under which the oil is used greatly affect the useable life and it is the operating conditions that can greatly accelerate the depletion of the additive package in the oil. When additive depletion takes place, it leaves the oil in a state of reduced protection of the equipment for whatever property the particular additive was designed to enhance. While there may be other purposes for additives, the primary purposes for gearbox oils are those added to inhibit rust, inhibit oxidation of the base oil, and improve either the anti-wear or the extreme pressure properties of the lubricant. Depletion of AW/EP additives or oil oxidation can be detected by oil analysis [51,52].

These and other additive depletions may be found in aging oil by oil chemical analysis detecting chemical elements (wear particles) in the oil. For example, AW and EP additives produce molybdenum (Mo), phosphorus (P), and zinc (Zn), anti-oxidants may produce boron (B) and Zn, corrosion inhibitors may be a source of Mo, Barium (Ba), Zn, and P, and so on [53].

Lack of Lubricant. Cylindrical roller bearings do not always have a separating lubricant film built up between the contacting rollers due to opposing surface speed. Thin film or even mixed lubrication under high roller-to-roller contact pressures leads to metallic contact between neighboring rollers, and this then increases friction, which consequently can lead to smearing and surface destruction. Micropitting can form on surface-hardened gears within the first several hours of operation if the gear box is not properly lubricated. The result is reduced gear tooth accuracy.

Water Contamination. Water contamination has a significant impact on wind turbine performance, particularly in offshore environments where water exposure is far greater. Water contamination can also cause the formation of rust on internal components. When water is present in oil it can cause additive depletion, stable emulsions, and higher viscosity. It can also lead to equipment issues such as filter blockage and accelerated wear of system components. Loss of oil additives or additive drop out occurs when additives present in the oil have a strong affinity for water. These additive molecules form bonds with water molecules and contaminate the whole lubrication system.

Microbial Growth. Micro-organisms such as bacteria, yeasts, and algae can grow and multiply inside lubricants if they encounter desired growing conditions such as water, optimum temperature (ranging from 15°C–52°C [60°F–125°F]), organic materials (food), and oxidative particles. Many strains of bacteria and molds will metabolize gear oil. If free water is present along with appropriate temperatures bacteria and molds will grow fast. Consequences include accumulation of acids, promoting corrosion since free water is present, as well as the formation of biological slimes that foul flow passages and moving parts. Microbial colonization of lubricants is also associated with fetid odors, asthma, and skin allergies. The organisms can quickly deplete oil quality and performance [54].

Foaming. Foaming is another notable equipment challenge for wind turbine operators. For example, when foam bubbles up and breaks through a shaft seal it makes a mess inside the nacelle, creating a "slip" safety hazard. Further, as foam forms on the surface of the oil it may interfere with the oil level float switch, giving a false reading and causing a potential alarm. Finally, if foam enters the oil circuit a momentary loss of oil pressure or flow could occur, also giving rise to an alarm. All instances could result in unnecessary down time [55].

3.4.2 Techniques to Mitigate a Negative Lubricant Role in Gearboxes and Bearing Failures

High-quality formulations are designed to protect equipment from common issues such as scuffing, micropitting fatigue, and rust and corrosion. A proper formulation also ensures protection at extremely high temperatures and good reliability at low temperatures. Wind turbine operators should use a lubricant with a balanced formulation, including the right mix of advanced base oils and additives, to ensure long-lasting performance.

Lubricants formulated with specific additives can help mitigate various negative effects of lubricants on performance of gearboxes and bearings [55,56].

For example, using lubricants with additives improving the oil's resistance to water contamination also improves its wet oil filterability. High-quality lubricants can be formulated with specific base stocks and additives that are engineered to resist foaming, helping to reduce related concerns.

Some kinds of additives enhance the oxidation film on the fresh metal surface [57]. The effect of extended bearing life is more likely due to the oxidation film formed by a tribochemical reaction. Oxidation film can prevent a fresh metal surface from being exposed to lubricants and keep the raceway surface chemically stable.

To mitigate formation of micropitting effects, operators should select an oil formulated with a micropitting additive package, such as conventional extreme pressure additives, as well as employing a gear finish as specified by American Gear Manufacturers Association's AGMA 6006 standard.

To combat the foam-causing effects of the gear oil additive package, as well as the effects of water and contaminants, all industrial gear lubricants contain foam-control additives. Some suppliers use the term "antifoam" for additives that keep foam from forming and additives that help collapse foam after it has formed. To be effective, the foam-control additive must be fully dispersed in the oil [58].

An oil formulated with advanced base fluids that provide a high viscosity index—generally 160 or higher—and a lower traction coefficient can also help. The higher viscosity index can provide a thicker lubricant film at operating temperature, and the lower traction coefficient can help increase energy efficiency.

3.4.3 Oil Analysis Identifying the Cause of the Failure

Gear oil contamination due to particles, either introduced during manufacturing or internally generated, can deteriorate the film formation capability of the lubricant. Debris travelling with lubricant flow can damage the surface of gearbox components via erosive wear. These particles, called *wear debris,* mix with the lubricating oil. Important information on failure cause of gearboxes and bearings may be obtained by performing wear debris analysis and online oil analysis [52,59].

By examining the wear debris, one can find out about the current status of the gearbox. Wear debris analysis is undertaken in different ways, but generally the particles are categorized in terms of their quantity or concentration, size, morphology, and composition. The associated wear characteristics are the severity, rate, mode, and source of the wear.

The oil analysis can be utilized as a part of a proactive maintenance strategy, but wear debris analysis can only be used to monitor active primary wear. Examples of typical wear debris produced by machines with rolling bearings and gear teeth, which undergo a nonconformal, rolling–sliding type of contact situation, are ferrous particles of varying shapes and sizes of between 10 and 1000 μm.

3.4.4 Practice of Top-Treating Oil with Additives

A new, trending technique to extend oil life is top-treating wind turbine oils with additives. The concept is simple: operators use condition monitoring to identify when an additives in oil starts to deplete and then re-additize the oil with the addition of after-market additive packages [60].

There are two main challenges: the potential for an unbalanced formulation and increased safety concerns. Additive top treating may actually introduce new components, or contaminants, that could impact the performance of wind turbine equipment by generating an unbalanced formulation. For example, surface-active additives like anti-wear additives and rust inhibitors both compete for space on the metal surfaces in a gearbox. Formulating an oil so that both of these types of additives are present in the correct amounts to properly protect the machine elements from both wear and rust is a delicate balance. Topping up with different ratios or different types of these additives could disturb the balance, creating more potential harm than good.

Top-treating oils with additives also increases how often you have to interact with your equipment. While top treating may not be as invasive as flushing and replacing an oil, regular additive top treating requires more frequent equipment interaction, which in turn increases the potential for safety issues.

However, this practice is not recommended since the key to long-term lubrication performance is to select an oil that is formulated with the right mix of advanced base oils and additives to deliver the required performance over many years without the need for additive top treating. Additionally, lubricant suppliers that embrace this balanced formulation approach stand behind their products. For example, some of ExxonMobil's synthetic wind turbine oils are warrantied for up to seven years, demonstrating the capability for the fluid to protect the machine even after 60,000 hours in service [56].

REFERENCES

1. Guerzoni F. 2013. Challenging lubrication application, *Wind Power International*, May. http://www.windpowerinternational.com/features/featureworld-wind-technology-shell-lubricants-dr-felix-guerzoni-bearing-grease/
2. Van Rensselar, J. 2013. Extending wind turbine gearbox life with lubricants, *Tribology and Lubrication Technology*, Vol. 69, Issue 5: 40–49. http://callcenterinfo.tmcnet.com/news/2013/05/25/7161084.htm
3. Holm A-P. 2009. Specialty lubricants for optimum operation, *Wind Systems Magazine*, September-October issue: 42–47, Online. Available: http://windsystemsmag.com/article/detail/23/specialty-lubricants-for-optimum-operation
4. Lugt Piet M. 2012. Innovative bearing technology for wind turbines, SKF Engineering & Research Center, November. https://www.ltu.se/cms_fs/1.99723!/file/Lugt-Tribodays%202012%20Wind%20Turbines.pdf
5. Chudnovsky B. H. 2017. *Transmission, Distribution, and Renewable Energy Generation Power Equipment: Aging and Life Extension Techniques*, Second Edition, Boca Raton: CRC Press/Taylor & Francis Group. https://www.crcpress.com/Transmission-Distribution-and-Renewable-Energy-Generation-Power-Equipment/Chudnovsky/p/book/9781498754750
6. Braun H. R. 2011. Wind Turbine Grease Lubrication, NREL Wind Turbine Tribology Seminar, Broomfield, Colorado, November 15–17: 45 http://www.nrel.gov/wind/pdfs/day1_sessionii_3_exxonmobil_braun.pdf
7. Key Wind Turbine Lubrication Points. Lubricants Improve Efficiency and Longevity of Wind Turbines, 2013. Dow Corning. http://www.azom.com/article.aspx?ArticleID=4891

8. Moore M. 2011. Lubricating wind components, *Wind Systems Magazine*, September issue: 52–57. http://windsystemsmag.com/media/pdfs/Articles/2011_Sept/0911_Shermco.pdf

9. Lubrication technology and increasing the reliability of wind turbine components, 2011. Dow Corning Corporation, Form No. 80-3682-01. http://www.dowcorning.com/content/publishedlit/80-3682.pdf

10. Specialty Lubricants for the Wind Energy Industry, 2010. Dow Corning Corporation, Form No. 80-3452A-01. http://www.dowcorning.com/content/publishedlit/80-3452.pdf

11. Martino J. 2013. Wind Turbine Lubrication and Maintenance: Protecting Investments in Renewable Energy, *Power Engineering Magazine*, vol. 117, issue 5. https://www.power-eng.com/articles/print/volume-117/issue-5/features/wind-turbine-lubrication-and-maintenance-protecting-investments-.html

12. Blade Bearing Lubrication in Windmills, Dow Corning Internal. www.edn.com/Pdf/ViewPdf?contentItemId=4137370

13. Standard Test Method for Determination of Corrosion-Preventive Properties of Lubricating Greases Under Dynamic Wet Conditions (Emcor Test), SKF Grease Test Rig EMCORO. Available: http://www.skf.com/binary/79296820/EMCOR.pdf

14. Guerzoni F. 2014. Lubricants' role in blade performance, *Renewable Energy Focus Magazine*, January/February issue. http://www.renewableenergyfocus.com/view/37563/lubricants-role-in-blade-performance/

15. Mooney G. 2016. Facing the Arctic challenge at the world's most northerly wind farm, Linkedin, *Pulse*, October 18. https://www.linkedin.com/pulse/facing-arctic-challenge-worlds-most-northerly-wind-farm-gavin-mooney/

16. Diemand, D. 1990. Lubricant at Low Temperatures. CREEL Technical Digest TD 90-01, Cold Regions Research & Engineering Laboratory, Hanover, New Hampshire. www.dtic.mil/dtic/tr/fulltext/u2/a234536.pdf

17. Lacroix A., Manwell J. F. 2000. Wind Energy: Cold Weather Issues, University of Massachusetts at Amherst, Renewable Energy Research Laboratory, June issue: 17. https://pdfs.semanticscholar.org/a66a/4943bdd3cb4e20b1ae451433bf560c6c5b16.pdf

18. Hansen M.O.L. 2008. Experience with Wind Turbines in Arctic Environment, Sustainable Energy Supply in the Arctic Conference, Sisimiut, Greenland, March.

19. Laakso T., Holttinen H., Ronsten G., et al. 2003. State-of-the-art of Wind Energy in Cold Climate, Finland, April. https://www.vtt.fi/inf/pdf/workingpapers/2010/W152.pdf

20. It's Very Cold – So What? Technical Article, Kluber Lubrication. https://www.klueber.com/ecomaXL/get_blob.php?name=Its_Very_Cold_-_So_What-L.pdf

21. Zgoul M. and El-Thalji I. 2011. Wind Energy Operation and Maintenance Practices in Desert Climate: threats, challenges and solutions, Global Conference on Renewable Energy and Energy Efficiency for Desert Regions (GCREEDER 2011), Amman-Jordan, April.

22. Leather J. 2011. Fundamentals of Lubrication Gear Oil Formulation, US Department of Energy, National Renewable Energy Laboratory (NREL), Wind Turbine Tribology Seminar, A Recap, Broomfield, Colorado, USA, November 15–17: 8–9. http://www.nrel.gov/docs/fy12osti/53754.pdf

23. Barr D. 2002. Modern Wind Turbines: A Lubrication Challenge, *Machinery Lubrication*, September. https://www.machinerylubrication.com/Read/395/wind-turbine-lubrication

24. Siebert H. 2006. Wind Turbines Power Up with Oil, *Lubrication & Fluid Power*, September–October Issue. http://www.klubersolutions.com/pdfs/_misc/wind_turbines_power_up_with_oil.pdf

25. Daschner E. 2011. Wind Turbine Lubrication Challenges and Increased Reliability, MDA Forum, Hannover Messe: 15. http://files.messe.de/abstracts/43459_7_Daschner_ExxonMobile(1).pdf

26. Standard DIN 51 517 "Lubricants – Lubricating Oils" - Part 1: Lubricating oils C, Minimum Requirements, PART 2: Lubricating oils CL, Minimum Requirements; PART 3: Lubricating oils CLP, Minimum Requirements, Publisher Deutsches Institut fur Normung E.V. (DIN), Revision / Edition: 14, Date: 02/00/14, Germany.

27. Hoehn B-R., Oster P., Tobie T., Michaelis K. 2008. Test Methods for Gear Lubricants, GOMABN, *Goriva I Maziva (Fuels and Lubricants)*, v. 47, No. 2: 129–152. https://hrcak.srce.hr/file/37724

28. Focus on Wind Turbines, Castrol: 8. http://www.trademarkoil.com/PDF/Castrol%20wind_turbines_brochure_EN.pdf29. MiCRoLuBGeaR© Lubrication device for pitch gears, User Manual, Grupo Técnico RIVI – Lincoln Spain: Lube systems & equipment. http://www.rivi.net/aplicaciones/energias-renovables-aerogeneradores/microlugear-v2.pdf

30. Optimizing wind farm performance, 2018. SKF-Lincoln, PUB LS/S2 14161/1 EN July: 20. http://www.skf.com/binary/30-141953/14161EN.pdf

31. Morling R. 2013. Remote solution for automatic lubrication of wind turbine bearing systems, *Design Solutions*, Nov 27. https://www.connectingindustry.com/DesignSolutions/remote-solution-for-automatic-lubrication-of-wind-turbine-bearing-systems.aspx

32. Cigno E. 2016. Tribological Study of High Performance Bio-Lubricants Enhanced with Ionic Liquids for Use in Wind Turbines, Theses, Rochester Institute of Technology. scholarworks.rit.edu/cgi/viewcontent.cgi?article=10193&context=theses

33. Dvorak P. 2016. RSC Bio Solutions expands series of biodegradable lithium complex greases, *Wind Power Engineering*, March. https://www.windpowerengineering.com/operations-maintenance/lubricants/rsc-bio-solutions-expands-series-biodegradable-lithium-complex-greases/

34. Non-toxic, biodegradable and renewable lubricants for Wind Turbines. http://www.sotaventogalicia.com/en/projects/non-toxic-biodegradable-and-renewable-lubricants-for-wind-turbines

35. Wind Turbine Reliability – Understanding and Minimizing Wind Turbine Operation and Maintenance Costs, 2004. Sandia National Laboratory, SAND2004-5924. https://prod-ng.sandia.gov/techlib-noauth/access-control.cgi/2004/045924.pdf

36. Skills Standards for Wind Turbine Technicians, Washington State Center of Excellence for Energy Technology. http://cleanenergyexcellence.org/wpcontent/files_mf/windstandards86.pdf

37. Wind Turbine Technician Skill Standards: 21. http://tssb.org/sites/default/files/wwwpages/repos/pdfiles/WindTurbineTechSS.pdf

38. Hennigan G. 2018, Maximizing Maintenance Dollars in a Post-PTS World, *H&N Wind*, February issue.
39. Muller J., Errichello R. 2002. Oil Cleanliness in Wind Turbine Gearboxes, *Machinery Lubrication*, July/Aug issue: 34–40. http://www.machinerylubrication.com/Read/369/wind-turbine-gearboxes-oil
40. Sayles R. and Macpherson P. 1982. Influence of Wear Debris on Rolling Contact Fatigue. Rolling Contact Fatigue of Bearing Steels. ASTM STP 771: 255–274.
41. ANSI/AGMA/AWEA 6006-A03 "Standard for design and specification of gearboxes for wind turbines; Section 6 – Lubrication". 2004. American Gear Manufacturer's Association (reaffirmed June 29, 2016).
42. ISO Standard 4406:99 Hydraulic fluid power – Fluids – Method for coding the level of contamination by solid particles, International Organization for Standardization, 12/01/1999.
43. Barrett M. P., Stover J. 2013. Understanding Oil Analysis: How It Can Improve Reliability of Wind Turbine Gearboxes, *Gear Technology*, November/December issue: 102–109. http://www.geartechnology.com/articles/1113/Understanding_Oil_Analysis:_How_it_Can_Improve_Reliability_of_Wind_Turbine_Gearboxes
44. Cummins J. 2013. Wind Turbine Lubrication, Hydrotex Lubrication University, STLE Houston Chapter Meeting, September 11: 66. http://www.stlehouston.com/2HoustonSTLE/2013-2014/Program/Wind%20Turbine%20Lubrication%20STLE%20Sept%20Meeting%202013%20Rev.pdf
45. CJC™ Fine Filter Technology. 2014. Karberg & Hennemann GmbH & Co. KG. http://www.kinnearspecialties.com/downloads/Fine_Filters.pdf
46. Barnes M. and Dvorak P. 2015. Contamination control for wind turbine gearboxes, *Wind Power Engineering*, March issue. http://www.windpowerengineering.com/featured/business-news-projects/contamination-control-for-wind-turbine-gearboxes/
47. Lumley S.L-L. Oil analysis in wind turbine gearbox, 2914. *Energize*, July issue: 65–71. www.ee.co.za/wp-content/uploads/2014/07/energize-july-2014-pgs-65-71.pdf
48. Han B., Binns J., Nedelc I. 2016. In Situ Detection of Hydrogen Uptake from Lubricated Rubbing Contacts, *Tribology Online*, vol. 11, no. 2: 450–454. https://www.researchgate.net/publication/301746471_In_Situ_Detection_of_Hydrogen_Uptake_from_Lubricated_Rubbing_Contacts/download
49. Kohara M., Kawamura T., Egami M. 2006. Study on mechanism of hydrogen generation from lubricants, *Tribology Transactions*, 49: 53–60.
50. Lu R., Nanao H., Kobayashi K. et al. 2010. Effect of lubricant additives on tribochemical decomposition of hydrocarbon oil on nascent steel surfaces, *Journal of the Japan Petroleum Institute*, 53(1): 55–60.
51. Lutz G. A., Jungk M., Lauer B. 2012. Full Life Wind Turbine Gearbox Lubricating Fluids, Final Report, Submitted by Dow Corning Corporation to United States Dept. of Energy (DOE), February: 111. https://www.osti.gov/servlets/purl/1041556/
52. Coronado D. and Wenske J. 2018. Monitoring the Oil of Wind-Turbine Gearboxes: Main Degradation Indicators and Detection Methods, *Machines*, vol. 6, no. 25, 20. https://www.mdpi.com/2075-1702/6/2/25/pdf

53. Sihna, Y., Steel, J. A., Andrawus J.A., Gibson K. 2015. Significance of effective Lubrication in mitigating system failures – a wind turbine gearbox case study, *Wind Engineering*, vol. 38, No. 4: 441–450. http://openair.rgu.ac.uk
54. Singh H. 2017. Wind Turbine Tribology, *The South African Mechanical Engineer*, SA Institute of Tribology, vol. 67, April: 28–29. http://79.170.40.230/sait.org.za/rw_common/plugins/stacks/armadillo/media/SAME_April2017.pdf
55. Dvorak P., Rojas L. 2016. Why lubricant formulation matters, *Wind Power Engineering*, February. https://www.windpowerengineering.com/operations-maintenance/lubricants/why-lubricant-formulation-matters/
56. Froes M. 2018. Mitigating turbine downtime with proper lubricant care, *Wind Power Engineering*, February. https://www.windpowerengineering.com/mechanical/gearboxes/mitigating-turbine-downtime-proper-lubricant-care/
57. Uyama H. and Yamada H. 2013. White Structure Flaking in Rolling Bearings for Wind Turbine Gearboxes, American Gear Manufacturer Association (AGMA), Paper 13FTM15. http://www.nrel.gov/wind/pdfs/day2_sessioniv_1_nsk_uyama.pdf
58. Beercheck D. 2007. Controlling Foam in Industrial Gear Oils, *LUBES'N'GREASES Magazine*, August.
59. Ribrant J. 2006. Reliability performance and maintenance – A survey of failures in wind power systems, Master Thesis written at KTH School of Electrical Engineering, Finland, http://faculty.mu.edu.sa/public/uploads/1337955836.6401XR-EE-EEK_2006_009.pdf
60. Hennigan G. 2016. Maintenance and Lubrication of Wind Turbines, *Alternative Energy Magazine*, July issue. https://www.altenergymag.com/article/2016/07/maintenance-and-lubrication-of-wind-turbines/24074

Lubrication Products in Electrical Industry

4

4.1 LUBRICANTS FOR POWER ELECTRICAL EQUIPMENT

A very important issue of best performance and long useful life for modern mechanical and electrical equipment is which lubricants are used during manufacturing and maintenance. Wrong lubrication technique, interval, or choice of lubricant can result in equipment failure. Lubricant manufacturers often publish the results of detail and thorough studies focused on the lubrication properties of the products they manufacture. The studies and comments on lubrication products for electrical apparatus in different conditions and environments are summarized in this chapter. It is very helpful to learn what they recommend for specific applications.

4.1.1 Dow Corning Corporation (USA)

4.1.1.1 Proper lubricant selection

In many applications in circuit breakers or any other electromechanical device requiring lubrication, conventional or older-technology lubricants continue to be used. However, highly engineered lubricants would deliver much better performance. Not understanding the features and benefits of engineered lubricants can lead to decreased reliability of electrical apparatus. It been demonstrated that modern engineered lubricants represent an opportunity to provide reliable equipment operation and increase service life for switchgear and circuit breakers. Modern lubricants are able to outperform conventional lubricants, excel at temperature extremes, resist drying out, and help prevent wear and corrosion that may lead to equipment seizure.

By paying attention to application details and considering the goals and objectives, a proper lubricant selection for any given application may be achieved. Factors influencing lubricant selection include component type (gear, bearing, etc.), kinds of surfaces in contact (metal-to-metal, metal-to-plastic, etc.), their relative motion, and temperature. Goals or objectives may include extended lubrication intervals, increased reliability, elimination of corrosion, improving trip time, and many others.

Knowledge of modern lubricant technology can be put to best use to achieve one's objectives once application details are understood and goals are prioritized. Four key factors to consider are: load, environment, temperature, and speed. After that end-users and designers can begin to make decisions on how to properly lubricate any particular application [1].

For example, devices that move frequently and operate under light to moderate loads without reaching temperature extremes can be successfully lubricated with liquid-based lubricants. On the other hand, as speed or frequency of motion decline and loads or vibration increase and temperatures reach extremes, lubricants where lubricating solids are present may be used, such as grease fortified with molybdenum disulfide or graphite or other lubricant forms such as lubricating paste.

4.1.1.2 Lubrication of outdoor electrical equipment

Outdoor electrical equipment is often exposed to the environment leading to accelerated aging of circuit breakers, disconnects, and switchgear. Poor circuit breaker and disconnect reliability are often caused by seizure triggered by rust and galvanic corrosion. Metal surfaces of outdoor electrical equipment are often exposed to oxygen, humidity, and fly ash, which results in not only environmental corrosion but also degradation of lubricants, resulting in seizure. Reliability of outdoor electrical equipment may in some cases be improved through careful lubricant selection.

Metal-to-metal contact in equipment exposed to environmental elements and static conditions for long periods of time may be best served by applying lubricating paste or dry film lubricants rather than grease. This choice is based on the ability of solids to provide a more robust, continuous protective film that lasts longer than films provided by liquid-based lubricants. Applications such as yoke adjustments, ball and socket connections, pins, bushings, and some gears and cams may be able to operate more freely over time if lubricated with solid-based lubricants [1].

There are other ways for the lubricant to degrade in outdoor electrical equipment: grease often degrades via oil evaporation and oil separation. *Loss of base oil* to evaporation leads to increasingly stiff grease that may in time impact trip times. Mineral oil–based grease tends to be more prone to evaporation than synthetic greases.

Static conditions may lead to *oil separation* and loss due to the effect of gravity. Thickeners have different densities than base oils. Gravity acts to separate them from each other. Thickeners are designed to release oil over time and they do not chemically bind with oil molecules. Most thickeners hold oil through capillary action or physical entanglement. The pull of gravity may overcome those things over time.

> *Oxidation* is a form of degradation that is more likely to occur in industrial applications where operating temperatures exceed 70°C (160°F) on a continuous basis.
>
> *Solvent washout* can occur when penetrating oils are used to maintain an equipment. Most lubricants are hydrocarbon based. Solvents in penetrating oils are hydrocarbon based also, and as chemically similar compounds they tend to dissolve each other quite well.

Studies and long field experience showed that *fluorosilicone*-based grease offers excellent resistance to evaporation and to hydrocarbon-based solvents and also resists washout by penetrating oils.

Equipment manufacturers and end-users can affect equipment reliability by choice of lubricant and selected or suggested lubrication interval. Increased reliability may come from use of lubricants that resist degradation over time and that are properly applied to components to protect them from wear and corrosion.

4.1.1.3 Molykote® and Dow Corning brand lubricants for power equipment

In electric power generation and distribution, high-voltage circuit breakers and disconnect switches are expected to remain in service for up to 20 years without major service. The temperature inside a circuit breaker may range from −40°C to 60°C (−40°F to 140°F) over the course of a year with wide daily fluctuations during the same season. Power equipment is also often exposed to various environmental extremes such as severe cold, extreme heat, wind, water, and airborne contamination. In such conditions, conventional mineral oil–based lubricants are often unable to function properly.

Mineral oil–based lubricants are more prone to drying out when exposed to high temperatures due to their increased volatility compared to the more thermally stable synthetics. Mineral oils also tend to have low viscosity indexes. They tend to thicken quickly as temperature decreases and thin quickly when temperature increases. In the case of a trip coil bearing, use of the wrong lubricant in colder conditions may mean the difference between an acceptable trip versus a delayed or failed trip. Over time, mineral oil–based grease may also dry out enough to prevent a successful trip.

In the electrical industry, synthetic lubricants such as *silicone- and polyalphaolefin (PAO)-based* lubricants provide longer service life due to their ability to resist changes in viscosity and, like a fluorosilicone grease, have been demonstrated to provide many years of successful service without drying out. The engineered properties of synthetics serve well in a wide range of applications eliminating seasonal lubricant change-outs. Synthetics are widely used by original equipment manufacturers as life lubricants.

Most high voltage equipment today is constructed without grease ports, making it very difficult to re-lubricate unless it is brought into the shop and disassembled. Because of this, maintenance personnel should use synthetic lubricants to take advantage of longer service life and better performance at temperature extremes [2].

Molykote® and Dow Corning® brand lubricants can provide proven long-term, reliable lubrication over a wide service temperature range and in a variety of environments. Fluorosilicones in particular resist solvents such as penetrating oil, and they do not dry out and harden, even after many years of exposure to the elements and static conditions. Formulated to address specific requirements, Molykote® and Dow Corning® lubricants can help reduce seizure and increase circuit breaker performance and reliability. They may allow extended maintenance intervals and reduced lifecycle cost. The lubricants may also generate increased customer satisfaction by reducing equipment failures.

An example of such products is Molykote® *3451 Chemical Resistant Bearing Grease.* This fluorosilicone grease provides superior resistance to most chemicals and can be used at high temperatures with service temperature range from −40°C to 232°C (−40°F to 450°F) and with heavy loads. This grease combines fluorosilicone oil with PTFE thickener. It provides exceptional lubrication on needle bearings in circuit breakers and may be used in roller and sleeve bearings, silicone o-rings, and gaskets, etc. This grease reduces wear and corrosion, resists drying out; is solvent resistant, and chemically inert.

Another Molykote® product is *33 Extreme Low Temperature Bearing Grease,* which delivers exceptional performance in cold environments.

Its service temperature range is −73°C to 180°C (−100°F to 356°F). It lubricates trip latch and coil bearings, antifriction bearings, and plastic and rubber parts under light to moderate loads. Molykote® 33 grease is resistant to oxidation and moisture, and it is compatible with many plastics and elastomers. It is based on phenylmethyl silicone oil and lithium soap thickener, it may be used on gaskets in presence of SF_6, on flange surfaces, it resists drying out, it is solvent resistant and chemically inert, and it may enable longer service life.

Original equipment manufacturers may apply different greases, including Dow Corning grease; however, a thorough testing of the equipment is always required for each specific lubrication product application to be accepted as OEM recommendation and included into OEM equipment manuals.

Some of the most useful lubricants for electrical applications manufactured and tested by Dow Corning are presented in Refs. [3,4,5]. These brochures list composition, service temperature range, potential applications and benefits and features of Molykote® brand lubricants for multiple application in electrical industry. Collected in Table 4.1 are some most popular Dow Corning/Molykote lubricants, pastes, and sealants widely used in the electrical power industry.

4.1.2 Nye Lubricants (USA)

Nye offers the highest performance synthetic lubricants that meet today's OEM design and specification requirements, involving both new equipment designs as well as upgrading existing equipment currently in operation. Nye synthetic lubricant formulations not only minimize component friction and wear to ensure free mechanical operation, but also offer superior protection from aggressive environments, preventing component corrosion/oxidation that could otherwise develop into electrical resistivity problems over time [6].

Shown in Table 4.2 are some Nye lubricants for electrical and mechanical applications in the power transmission and distribution industry. The table shows composition (base oil and thickener type), temperature (T) working range, and typical applications.

> *Mechanical components* must remain well lubricated to provide low friction and free operating motion, even with long periods of infrequent operation. Nye's synthetic formulations offer extended long-term mechanical component protection from oxidation and environmental degradation, with superior protection from long-term exposure to UV, moisture, and water spray, even while exposed to wide temperature swings.

TABLE 4.1 Dow Corning/Molykote lubricants, pastes, and sealants for electrical industry

LUBRICANT (COLOR)	COMPOSITION: BASE OIL, THICKENER	WORKING T RANGE °C (°F)	APPLICATION
Molykote™ BG-20 grease (beige)	PAO, Li complex, NLGI grade 2, biodegradable	−46 to 182 (−50 to 380)	Plain, ball, and sleeve bearings, heavy load, current-carrying parts.
Molykote™ 3451 Chemical Resistant Bearing Grease (white)	Fluorosilicone, PTFE, NLGI grade 2	−40 to 232 (−40 to 450)	Roller, needle, and sleeve bearings, silicone o-rings and gaskets. Suitable in harsh environments with chemicals, acids, and alkalis.
Molykote™ 33 Extreme Low T Bearing Grease (pink/gray)	Phenylmethyl Silicone oil, Li soap, NLGI grade 1.5-2	−73 to 180 (−100 to 400)	Ball bearings under light to moderate loads Applicable where cold T may hinder movement in trip latch mechanism, disconnect contact surfaces
Molykote™ BR-2 Plus (black)	Heavy paraffinic and naphthenic mineral oils, Li soap, MoS2, NLGI grade 2	−29 to 129 (−20 to 262)	Latches, motor and gear wheels, moving surfaces, shafts, roller and plain bearings, bushings and sleeves, guides and tracks
Molykote™ 557 (colorless)	Silicone dry film in Stoddard solvent, aerosol	−40 to 43 (−40 to 109)	Ground surfaces, sliding surfaces, moving parts
Molykote™ G-n Metal Assembly Paste (black)	MoS2 complex and white solid lubricants in mineral oil	−18 to 399 (0 to 750)	Needle and roller pins, cams and bushings, metal-to-metal sliding surface contacts, gears and pinions, bevel gears, jack screws associated with racking mechanisms.
Dow Corning 832 Multi Surface Adhesive Sealant	Silicone RTV, neutral cure 100%	−55 to 149 (−67 to 300)	Sealing flange joints, bolts and nuts on breakers exposed to the weather
Dow Corning 111 Valve Lubricant and Sealant	Dimethyl silicone oil, silica	−40 to 204 (−40 to 400)	Gaskets (air-to air), non-silicone o-rings, sliding seals and surfaces
Molykote™ 1000 Solid Lubricant Paste	Mineral oil, thickener, powdered copper, graphite, white solids	−30 to 650 (−22 to 1202)	Bolted joints subjected to high T and corrosive environment, enabling non-destructive dismantling or re-tightening
G-Rapid Plus Paste	Mineral oil, thickener, solid lubricants MoS2, graphite	−35 to 400 (−31 to 752)	Threaded bolted connections, tooth and worm gearing, fitting ball and roller bearings

Electrical sliding and mating contact surfaces must remain free of oxidation and corrosion that could increase resistivity and free of wear debris and other contaminants that could compromise surface finish and prevent proper interfacing. Nye's synthetic formulations offer excellent stay-in-place lubrication, good film strength to prevent wear (normal operation and fretting due to thermal cycling), as well as protecting surface integrity from exposure to environmental elements and subsequent oxidation over time.

Arcing electrical contacts see flash temperatures sufficiently higher than organic molecules can withstand. Therefore, Nye's synthetic lubricant technology, developed specifically for high voltage arcing applications, incorporates chemistries that will flash off in an innocuous manner, preserving conductivity of mating contact surfaces without leaving behind carbonized contaminants. The use of an inappropriate lubricant that does burn leaving carbon deposits can result in the buildup of a resistive layer and generation of heat.

4.1.2.1 Nye lubricants for separable electrical connectors

Lubricants reduce friction and ease mating. A thin film of lubricant can reduce mating force by as much as 80%, an important factor in connector assembly. For electronic connectors with dozens or even hundreds of pins, or for automotive connectors that are in hard-to-reach places, a low insertion force makes assembly more efficient and ensures solid connections. For gold-plated connectors, an effective lubricant reduces the potential for noble metal wear during mating and separation.

Nye Lubricants designs and manufactures many lubricants for separable electrical connectors. Nye's connector lubricants can be divided into two general classes: lubricants for noble metal connectors and tin/lead connectors. For noble metals, fluoroethers are the lubricants of choice. They withstand extreme temperatures and resist aggressive chemicals and solvents. Five- and six-ring polyphenyl ethers are also extremely stable in thin film and offer an excellent track record on gold-plated connectors. For tin/lead connectors, synthetic hydrocarbons provide excellent film strength, broad temperature serviceability, and protection against fretting corrosion.

Lubricant applied to connectors guard against corrosion and the effects of harsh environments. With gold-plated connectors, this means protection against substrate corrosion. Thin gold plating can be microscopically porous. A thin film of lubricant can seal the pores, prevent substrate attack, and assure low contact resistance.

TABLE 4.2 Nye lubricants for circuit breakers and disconnects

NYE LUBRICANT	COMPOSITION: BASE OIL, THICKENER	TEMPERATURE RANGE, °C (°F)	TYPICAL APPLICATION
Rheolube® 362HM	PAO, Lithium	−40 to 125 (−40 to 257)	Cams, sliding plastic surfaces where tackifiers provide stay-in-place lubrication on plastic components, mechanical linkages of switch gear
Rheolube® 362HT	PAO Lithium	−54 to 125 (−65 to 257)	Multifunction switches, bearings, sliding surfaces, electrical components
Rheolube® 368**	PAO, Lithium	−40 to 125 (−40 to 257)	Wide range of operating mechanisms
Rheolube® 368AX-1	PAO, Lithium	−20 to 125 (−4 to 257)	Slides & highly loaded gears, requiring low friction and wear
Rheolube® 375*	PAO, Lithium	−54 to 125 (−65 to 257)	Sliding contact surfaces (switch blades, etc.), low mechanical torque/forces at low T
Rheolube® 462	PAO, Lithium	−54 to 130 (−65 to 266)	Version of Nye 360 series with improved corrosion prevention and wear reduction
Rheolube® 716R	Ester, clay	−54 to 150 (−65 to 302)	Rolling element bearings operating over a wide T range where smooth, quiet operation is essential. Ideal for low temperature, low torque applications.
Rheolube® 716HT	PAO, Lithium	−54 to 175 (65 to 347)	Rolling element bearings operating over a wide T range
Rheolube® 731	Ester Lithium	−30 to 100 (−22 to 212)	Lubricant for arcing contacts
Rheolube® 737S	Ester, clay	−60 to 120 (−76 to 248)	Gear boxes, applications requiring low torque
Rheolube® 748LT	PAO, Lithium	−35 to 100 (−31 to 212)	Lubricant for arcing contacts
Rheolube® 789DM	Ester, clay	−40 to 150 (−40 to 302)	Gear trains, roller element bearings

Product	Composition	Temperature range	Application
RheoTemp™ 763G*	PAO/AN, urea	−54 to 175 (−65 to 347)	Sliding contact exposed to higher T
RheoTemp™761G	PAO/AN, urea	−40 to 175 (−40 to 347)	Stationary separable electrical connectors
RheoTemp™ 761AG	PAO/AN, urea	−40 to 175 (−40 to 347)	Stationary separable electrical connectors
NyoGel® 741A	Silicone, Lithium	−60 to 200 (−76 to 392)	Pistons (metal on rubber, metal on metal) requiring very slow sliding friction at a wide range of operating temperatures
NyoGel® 760G**	PAO, silica	−40 to 130 (−40 to 266)	Connections with higher loading forces (primary disconnects, etc.)
NyoGel® 782D***	Glycol, silica	−29 to 100 (−20 to 212)	Arcing contacts requiring no carbon forming
NyoGel® 718B *, **	PPE, silica	4 to 260 (39 to 500)	Slow speed bearings exposed to very high temperatures or radiation, gold-plated contacts
NyoGel® 774VLF	PAO	−50 to 120 (−58 to 248)	Heavily loaded operating at slower speeds
UniFlor™ 8623B**	PEPE, silica	−25 to 250 (13 to 482)	Medium–high voltage breakers, heavy loaded mechanisms exposed to very high T, stands up to the most aggressive chemical environments
UniFlor™ 8917	PEPE, Melamine Cyanurate	−70 to 225 (−94 to 437)	Chemical resistant, high-temperature lubricant for electrical connections
UniFlor™ 8511	PEPE, PTFE	−50 to 225 (−58 to 437)	Low load/low current applications connectors
UniFlor™ 8512	PEPE, PTFE	−50 to 225 (−58 to 437)	Low load/low current applications connectors

Note: Voltage Range (kV):
* LV<1 kV
** MV=1–72 kV
*** HV>72 kV

Lubricant also prevents fretting corrosion to which tin/lead connectors are also subjected. Fretting corrosion is the result of low amplitude vibration caused by thermal expansion and contraction or nearby motion, as from fans, motors, or merely opening and closing a cabinet door. Fretting corrosion continually exposes fresh layers of metal surface to oxidation. A lubricant film minimizes metal-to-metal contact during vibration, protecting the connector from metal wear.

Popular Nye lubricants are used for stationary separable electrical connectors. Some of them are presented in Table 4.3 [7]. All lubricants from this

TABLE 4.3 Nye lubricants for stationary separable electrical connectors

NYE LUBRICANTS	TEMPERATURE RANGE, °C (°F)	TIN/ LEAD	HIGH CURRENT	PLASTIC COMPATIBLE
Fluorocarbon Gel 813-1	−70 to 200 (−94 to 392)	+	+	+
NyeTact® 502J-20-UV*	−40 to 120 (−40 to 248)	+		+
NyeTact® 523V-2-UV*	20 to 250 (68 to 482)			
NyeTact® 561J-20-UV*	−40 to 175 (−40 to 347)	+	+	+
NyeTact® 561J-35-AG **	−40 to 175 (−40 to 347)	+	+	+
NyeTact® 570H-2	−20 to 225 (−4 to 437)			+
NyeTact® 570H-25-UV*	−20 to 225 (−4 to 437)			+
NyoGel® 760G	−40 to 135 (−40 to 275)	+	+	+
NyoSil M25 (Oil)	−70 to 200 (−94 to 392)	+	+	+
Rheotemp® 761AG**	−40 to 175 (−40 to 347)	+	+	+
Rheotemp® 761G	−40 to 175 (−40 to 347)	+	+	+
UniFlor™ 8511	−50 to 225 (−58 to 437)	+	+	+
UniFlor™ 8917	−70 to 225 (−94 to 437)	+	+	+

Source: Adapted from [7] Lubricants for Stationary Separable Electrical Connectors, NYE LubeNotes, No. 9, 2012

Note: * – UV Tracer allows easy identification of a lubricant on contact surfaces
  ** – Lubricant may be used on silver and silver-plated contacts

table are applicable to *low current* electrical connectors. All of them may be used on the connectors made of or plated with *noble metals*. Some of the lubricants contain so-called *UV tracer*, which is a fluorescent dye added to the lubricant. The dye will collect at all leak sites. When the system is inspected with a high-intensity ultraviolet (UV) or blue light lamp, the lamp will show the precise location of every leak with a bright glow. The dye can remain in the system indefinitely and does not affect system components or performance.

4.1.2.2 Nye lubricants for sliding contacts in electric switches

Grease for sliding electric switch contacts must meet similar demands as other mechanical sliding surfaces: film strength, appropriate low and high temperature fluid range, and stay-in-place capability.

A switch grease's ability to prevent wear is critical since two problems are created by wear debris: (a) when the contact is closed it can inhibit current flow increasing millivolt (mV) drop, and (b) when contact is open conductive wear debris can cause open circuit resistance (OCR) problems. In either case, switch performance is compromised [8].

The viscosity of the base oil in a switch grease should complement the contact force of the switch. Low current/low contact force applications require lighter base oils. High current/high contact force applications benefit from more viscous base oils. Lubricants recommended to use for sliding contacts are listed in Table 4.4 [9].

4.1.2.3 Nye lubricants for arcing contacts

The temperatures reached in an electric arc are sufficiently high to degrade any organic molecule. A tendency of a lubricant to "burn cleanly" is a definite advantage when applied to arcing contact. Greases for arcing contacts should be formulated with fluids and thickeners that degrade with fewer by-products than traditional greases. A recent innovation is the addition of an additive that scavenges surface oxides to reduce contact resistance. Lubricants recommended for arcing contacts are shown in Table 4.5.

Greases that oxidize under arcing conditions pose a special problem for low load/low current applications. Traditionally, cleaner burning glycols were used to minimize carbon buildup. A new approach to eliminating problems associated with oxidation is nonburning switch lubricant technology to which UniFlor™ nonburning perfluoropolyether-based greases are designed. Dispersed in a non-flammable, ozone-safe, fluorinated solvent, these greases leave a thin film of lubricant, ideal for low load/low current applications. As an

TABLE 4.4   Nye lubricants for sliding contacts in electric switches

NYE LUBRICANTS	RECOMMENDED CONTACT FORCE (g)[b]	CURRENT[c]	TEMPERATURE RANGE, °C (°F)	PROPERTIES[d]
Rheolube® 368	150	***	–40 to 125 (–40 to 257)	Y
Rheolube® 789DM[a]	80	*** and **	–40 to 150 (–40 to 302)	X
Rheolube® 362HT	50	* and **	–54 to 125 (–65 to 257)	Y
Instrument Grease 732C[a]	20	* and **	–54 to 150 (–65 to 302)	X
Rheolube® 716HT[a]	20	* and ** and ***	–54 to 175 (–65 to 347)	
Rheolube® 737S	20	*	–60 to 120 (–76 to 248)	X,Y

Source: Adapted from Lubricants for Sliding Contacts in Electric Switches, Nye Lubricants, LubeNotes. https://www.nyelubricants.com/site/get_document.php?id=3722
Note: [a] Use with caution around polycarbonate, ABS resins, Buna N, and other ester-vulnerable plastics and elastomers
   [b] Minimum contact force (g) that must be available for serviceability at –40°C
   [c] Current: *– low <1 amp, **– medium 1-10 amps, ***– high >10amps
   [d] X- Salt Water Resistant: Y- Plastic Compatible

additional benefit, this thin film does not attract dust and debris. An example is Nye grease UniFlor 8511, which is chemically resistant with excellent material compatibility and working T range from −50°C to 225°C (−58°F to 437°F).

4.1.2.4 Lubricants for distribution switchgear and other applications

Distribution switchgear may remain unactuated for long periods. In such devices the lubricants should play a protective as well as a lubricating role. They should not be oxidized over time and be water resistant and non-migrating. Because high temperatures may be induced by high current flow or high-temperature industrial conditions, wide temperature capability can be important.

The Nye greases recommended for distribution switchgear are Reolube® 362 HM (working T range −54°C to 125°C (−65°F to 227°F), Reolube® 368 and NyoGel® 760D, both with working T range from −40°C to 125°C (−40°F to 227°F). All three greases are plastic compatible. Actuator pistons and control valves may be lubricated with NyoGel 741A with silicone as base oil and soap thickener.

TABLE 4.5   Lubricants for arcing contacts

NYE GREASE	CONTACT FORCE[c]	TEMPERATURE RANGE, °C (°F)
Rheolube® 731[a,b]	**	−30 to 100 (−22 to 212)
Rheolube® 748LT[d]	**	−35 to 100 (−31 to 212)
NyoGel® 782G[a]	*	−40 to 100 (−40 to 212)
UniFlor™ 8512[d]	* and **	−50 to 225 (−58 to 437)

Source: Adapted from Lubricants for Sliding Contacts in Electric Switches, Nye Lubricants, LubeNotes. https://www.nyelubricants.com/site/get_document.php?id=3722

Note: [a] Use with caution around polycarbonate, ABS resins, Buna N, and other ester-vulnerable plastics and elastomers
   [b] New oxygen scavenging additive
   [c] Contact Force: * – Low (<100g*); ** – High (>250g**)
   [d] Plastic compatible

A number of Nye lubricants may be used for enclosures and electrical connections where a dielectric sealant is required. Reolube 368 could be used to coat gaskets and mating surfaces (boots, cases, etc.) to seal out environment. NyoGel 760G may be applied directly in and around connection and also to seal out environment. UniFlor 8917 is a very good chemical-resistant lubricant for electrical connections exposed to high temperature.

4.1.2.5 Anderol lubricants for contacts and switches manufactured by Nye Lubricants

Anderol was well known for manufacturing many lubricants accepted by in the industrial maintenance and electrical power management equipment industries as the products for use in circuit breakers and switches for many years. With the target to broaden and expand its product lines, Nye Lubricants is the official manufacturer of 6 of the Anderol® lubricants which have had their place in electrical equipment industries for many years.

Since 2011 Nye Lubricants is the official manufacturer of six of the Anderol® lubricants: Anderol® 732, Anderol® 752 Grade 1, Anderol® 752 Grade 2, Anderol® 757, Anderol® 786, and Anderol® 793 products. Nye is manufacturing these products according to the original process and control specifications as provided by Anderol [10]. The new Anderol greases manufactured by Nye Lubricants are presented in Table 4.6.

TABLE 4.6   Nye lubricant's family of Anderol® replacement lubricants

LUBRICANT (COLOR)	COMPOSITION: BASE OIL, THICKENER, NLGI GRADE, TEXTURE	WORKING T RANGE °C (°F), OIL SEPARATION (%)	DROPPING POINT, °C (°F), EVAPORATION (%)	PENETRATION WORKED (UNWORKED), mm/10	PERFORMANCE	APPLICATIONS
AND-732 (amber)	Diester, Li soap, 0, tacky	−40 to 150 (−40 to 302), 15	181 (358), 0.66	359 (397)	Friction and wear reduction, rust protection	Chain, gears, slides, bearings
AND-752-1 (tan)	PAO, Li complex, 1 smooth	−60 to 150 (−76 to 302), 3.7	296 (565), 0.20	315 (307)	Anti-wear, EP	Chain, gears, slides, bearings
AND-757 (amber)	Diester, Li soap, 1.5 tacky	−40 to 150 (−40 to 302), 1.4	192 (378)	295 (289)	Anti-wear, adhesion to metal	General purpose, electrical
AND-786 (black)	Diester, clay, 1.5 tacky	−20 to 150 (−4 to 302), 0.6	>300 (>572), 1.60	303 (301)	Corrosion/rust inhibited, load carrying at high speed	Gears, slides, bearings, pulleys
AND-793 (light brown)	Diester, Li soap, 2, smooth	−60 to 150 (−76 to 302), 4.8	192 (378), 0.60	289 (261)	High T stability	Fine precision instruments

Source: Adapted from Nye's Family of Anderol® Replacement Lubricants, NYE LubeNotes, February 2013

4.1.3 ExxonMobil (USA)

ExxonMobil manufactures many greases and oils widely used in electrical industry for multiple applications in circuit breakers, switches, and switchgears. Some of the most popular lubricants are presented in Table 4.7.

TABLE 4.7 ExxonMobil synthetic greases and oils for electrical industry

LUBRICANT (COLOR)	COMPOSITION: BASE OIL, THICKENER	NLGI GRADE	WORKING T RANGE °C (°F)	APPLICATION
Mobilgrease 28 (dark red)	PAO, organo-clay	1.5	−54 to 177 (−65 to 350)	Sliding surfaces, latch surfaces, gear, disconnect fingers, ball and roller bearings
Mobiltemp SHC 32 (dark red)	PAO, organo-clay	1.5	−54 to 177 (−65 to 350)	Sliding surfaces, latch surfaces, gear, disconnect fingers, higher water washout resistance
Mobilith® SHC 100 (red)	PAO, Li complex	2	−40 to 150 (−40 to 302)	Bearing, linkages, cams, gears, sliding parts, pins, pivots
Mobilith® SHC 007 (red), semifluid, pourable	PAO, Li complex	00	−40 to 150 (−40 to 302)	Bearing, linkages, cams, gears, sliding parts, pins, motors, pivots, heavily loaded applications
Beacon™ 325 (tan)	Ester, Li soap	2	−50 to 120 (−58 to 248)	Bearing and pivot pins, charging springs, sliding parts at vacuum interrupter, rotating parts
Mobil 1 Tri-Synthetic Oil, SAE grade 10W-30	Multigrade oil	–	down to −54 (−65)	Bearings, rollers, pins, wheels
Mobil 1 Tri-Synthetic Oil SAE grade 5W-30	Multigrade oil	–	down to −54 (−65)	Links, link gears

4.1.4 Cool-Amp (USA)

U.S.-based Cool-Amp has been family owned and operated by three generations since 1944. The company manufactures Cool-Amp, a silver plating powder, which is applied by hand and recommended for use on copper, brass, or bronze stationary parts. Cool-Amp can be used on buss bars, cable terminals, current transformer terminals, clamps and fittings, ham radios, PCB's, welders, or anywhere conductivity is needed on bolted copper, brass, or bronze parts. The powder reduces power loss and overheating, maximizes conductivity, assures long life and low maintenance; it also prevents oxidation and corrosion [11]. It is an alternative to silver plating/dipping of damaged electrical parts. The silver dipping process requires parts to be shipped for offsite dipping and shipped back before they can be used.

The company produces Conducto-Lube, a silver conductive lubricant for high amperage connections. There are no voltage limits in application. The grease is not soluble in water; however, it will gradually wear off if exposed to water for long periods of time. It does not change conductivity, but will conduct electricity and resistance does not change appreciably with temperature. Working temperature range is from −31°C to 210°C (−25°F to 410°F). Conducto-Lube contains 65%–75% of silver and 25%–35% of mineral oil. The silver used has particles that are 3–4 microns in size.

Conducto-Lube provides electrical and thermal conductivity, lubrication, and protection. The lubricant eliminates hot spots, increases conductivity, and prevents galling or pitting that will freeze joints. Conducto-Lube is successful in reducing resistance in hinge joint and knife blade switches due to arcing because the solids in the lubricant eventually imbed themselves into the minute imperfections of the contact surface resulting in a smooth surface.

4.1.5 Chemtronics (USA)

CircuitWorks Silver Conductive Grease is a silver-based grease utilizing an advanced silicone lubricant that is compatible with metal, rubber, and plastic. This grease is manufactured by ITW Chemtronics® [12]. It features high electrical conductivity, excellent thermal conductivity, and provides protection against wear, moisture, and corrosion. Because it is based on silicone oil, it remains stable in a wide temperature range from −57°C to 252°C (−70°F to 485°F). The silver/gray grease consistency is smooth paste; it has a very low viscosity vs. temperature change.

This silver conductive grease may be used for high and low power applications including lubrication of substation switches or circuit breakers and low or medium speed sliding contacts. Other usages could be for static grounding on seals or o-rings.

4.1.6 Sanchem (USA)

Sanchem manufactures NO-OX-ID Grease for corrosion protection of electrical contacts. The electrically conductive grease has been used in the power industry for over 75 years to prevent corrosion in electrical connectors from low micro-power electronics to high voltage switchgear. NO-OX-ID "A-Special" is the electrical contact grease of choice for new electrical installations and maintenance.

This grease keeps metals free from rust and corrosion from acids, salt, moisture, and various industrial chemical vapors in the environment. The grease prevents the formation of oxides, sulfides, and other corrosion deposits on copper, aluminum, and steel surfaces and conductors and lubricates the connection for easier maintenance. When this conductive paste is used on aluminum connectors in joints, NO-OX-ID A-Special prevents formation of oxide films, which cause high resistance and subsequent failures.

NO-OX-ID A-Special conductive grease is recommended by connector manufacturers for trouble-free joint connections. When nuts, mounting bolts, and cotter keys are coated with NO-OX-ID A-Special, they will never rust or freeze, assuring easy, trouble-free removal. NO-OX-ID A-Special should be used wherever the formation of a corrosive product will effect the proper functioning of the metal surface. This electrical contact grease is easily applied, easily removed, and gives long-lasting, reliable performance even on dissimilar metals. Some of the applications of this protective grease are: bolted connections, brackets, bus bar systems, cables and clamps connectors, and contact points in circuit breakers and switches [13].

4.1.7 Tyco Electronics (USA)

Tyco Electronics manufactures contact/joint sealants to protect electrical joints from galvanic and atmospheric corrosion. Corrosion damage of electrical joints is the strongest in heavy industry locations such as steelworks, chemical plants, refineries, etc., and also may occur in urban and rural areas.

Electric joins may be damaged by atmospheric corrosion caused by moisture and oxygen, while galvanic corrosion occurs when two dissimilar metals in the electrolytic series, for example, aluminum and copper, are in physical contact.

Galvanic problems are not present for aluminum-to-aluminum connectors, however, the oxide film forms rapidly on the surface of freshly cleaned aluminum exposed to air. This oxide film is an insulator and must be removed with a scratch brush in order to achieve a satisfactory and reliable electrical joint. The problem with aluminum is that the freshly cleaned surface will quickly re-oxidize, hence it is important to coat the surface with a contact sealant.

For Al to Cu connections, it is good practice to use contact sealant on the aluminum connector body and brushed into the strands of the aluminum conductor. Wherever possible, the aluminum conductor should be installed above the copper to prevent pitting from the galvanic action of copper salts washing over the aluminum connector and conductor when in a lower position.

Various sealant formulations have been developed to provide improved electrical and mechanical performance as well as environmental protection to the contact area [14]. The use of sealants is recommended for aluminum-to-aluminum or aluminum-to-copper connections. Sealants are also recommended for copper-to-copper joints that are subject to severe corrosive environments. Non-gritted sealants are recommended for flat connections and as a groove sealant in bolted connectors such as parallel groove clamps. Gritted sealant is primarily used in compression connectors. The sharp metallic grit particles provide multi-contact current-carrying bridges through remaining oxide films to ensure superior electrical conductivity.

TE Sealants with Fluoride. The joint sealants are mineral oil–based corrosion inhibitors with added chemicals to dissolve aluminum oxide. The sealants with added fluoride (drop point 65.6°C) are recommended for Al to Al and Al to Cu palm-to-palm joints, and for use on aluminum surface-to-surface bolted joints, such as bus-bar joints and terminal lugs. The sealant (EJC2) assists in breaking up the oxide film by chemically etching the connecting surfaces to ensure a low resistance joint.

TE Sealants with Lithium. Sealants with added lithium are recommended for bolted connections (Al to Al, Al to Cu, Cu to Cu) and on palm-to-palm joints (Cu to Cu). This sealant has a higher drop point (180°C) than the one with fluoride. Such a sealant (Alvania ALV300) seals the exposed surface to prevent re-oxidation and permanently excludes the future ingress of air and moisture. It is extremely adhesive, resistant to water, and has high temperature resistance to ensure continuous operation under all situations.

TE Sealants with Zinc. Joint compounds with added zinc particles have the highest drop point at 188°C; they provide excellent outdoor protection. Alminox ALM325G is recommended for compression joints and bolted connections (Al to Al and Al to Cu). Conductive zinc granules are suspended in a viscous petroleum oil base. Under pressure these granules make high-pressure contact points with the parent metal to provide a sound electrical connection, while the base material seals the joint to prevent further corrosion.

4.1.8 Hubbel/Burndy (USA)

The electrical joint protection compounds are manufactured by Burndy, currently a part of Hubbel Electrical Systems. Burndy's zinc-containing compounds called Penetrox™ have been used for many years in electrical industry. These products are oxide-inhibiting compounds producing low initial contact resistance, sealing out air and moisture, and thus preventing oxidation or corrosion. These pastes are usable over wide temperature ranges, providing a high conductivity, so-called "gas-tight-joint", or GTJ.

All Penetrox™ compounds contain homogeneously suspended particles. The particles assist in penetration thin oxide films, act as electrical "bridges" between conductor strands, aid in gripping conductor, improve electrical conductivity, and enhance integrity of the connection.

The specially formulated Penetrox™ compounds are for use with compression and bolted connectors providing an improved service life for both Cu and Al connections. Additionally, the nontoxic compounds are an excellent lubricant for threaded applications reducing galling and seizing.

4.1.9 Contralube (UK)

High voltage contact greases face a number of technical challenges. They must not only cope with the extreme temperatures that occur under arcing conditions but also minimize friction and withstand broad ambient temperatures without oxidizing, evaporating, or causing resistive failures; a high quality grease will also inhibit wear and corrosion, provide an element of environmental sealing, and control free motion.

Contralube brand consists of synthetic gels that are primarily used for the protection of electrical, electronic, data connections, and contact areas/surfaces. Composition of the lubricants is not disclosed. One of these products (Contralube 880) is a green gel that is primarily used as a contact surface lubricant and protector on high voltage/arcing contacts, circuit breakers and disconnects [15]. Typical applications for Contralube 880 gel include power distribution/switchboards, circuit breakers and disconnects, high voltage electrical substation contacts; power infrastructure, and distribution equipment.

Contralube 880 gel is used to prevent contact wear, corrosion, and the buildup of carbon deposits from arcing that can form a resistive layer that leads to overheating and a possible fire hazard. When arcing occurs across electrical contacts, the temperatures reached are sufficient to degrade any organic molecule and so a lubricant's ability to burn cleanly is a definite advantage; Contralube 880 gel is designed with a specialist formulation that burns away cleanly under arcing conditions, leaving no carbon deposits on the contact surface. It is synthetic product, therefore is cannot evaporate, it is water and UV resistant, and it does not contain silicone.

4.1.10 Electrolube (UK)

Electrolube, a division of H. K. Wentworth Limited, has been the leading supplier of contact lubricants since their invention by the founder in the 1950s. The lubricants increase the reliability and lifetime of all current-carrying metal interfaces, including switches, connectors, and busbars [16]. Some lubricants and their properties are shown in Table 4.8.

Electrolube has earned an unsurpassed reputation for the manufacture and supply of special lubricants to the automotive, military, aerospace, industrial, and domestic switch manufacturing sectors. The range has been developed over the years to accommodate many advances in these rapidly advancing industries; they combine excellent electrical properties and lubricity to improve movement and 'feel' characteristics, with plastics compatibility.

Electrolube products are electrically insulating in thick films, preventing tracking. In ultrathin films, i.e., between closed metal contacts, they allow the current flow; they also exhibit a neutral pH thereby avoiding surface corrosion.

4.1.11 Klüber (Germany)

Electrical contacts are normally held by plastic parts. Occasionally, rubber-elastic materials—also referred to as elastomers—can also be found in the vicinity of the contacts. The selection criteria for a lubricant depend on the parameters that are most important to be optimized. Some of these parameters are reduction of plug and unplug forces and resistance to fretting corrosion.

For lubricant selection, it is also of major importance to consider what kind of metal surfaces are moved against one another with what contact force. The adhesion of the lubricant depends on its chemical composition and consistency, but also on the contact material, the surface roughness, and the orientation of the roughness. As lubricants applied to live electrical contacts must withstand very high temperatures over a very long period of time, the

TABLE 4.8   Electrolube (UK) contact lubricants

LUBRICANT	COMPOSITION: BASE OIL, THICKENER, NLGI	TEMPERATURE RANGE, °C (°F)	APPLICATION
SGA	Complex Ester, clay, 1	−40 to 125 (−40 to 257)	Electrical fixed or moving contacts of contactors, busbars, knife switches, switchgear in heavy arcing conditions, and corrosive environments
SGB	Poly Alkylene Glycol, clay, 1	−35 to 130 (−31 to 266)	All types of electrical contacts and with most types of thermoplastics
CG53A	Poly Alkylene Glycol, Li complex, 1	−35 to 130 (−31 to 266)	Switches and connectors, high-voltage contacts in corrosive environments
CG60	PAO/Complex Ester, Li complex, 1	−45 to 130 (−49 to 266)	Column switches, rocker switches, and push-push switches in automotive and high quality domestic switch industries
EGF	PFPE, Fumed Silica, 2	−25 to 300 (−13 to 572)	Printed circuit edge connectors, plug connectors, rotary and sliding switches, gold contacts in aggressive environments
HVG	Poly Alkylene Glycol, Lithium Complex, 1	−35 to 130 (−31 to 266)	High voltage/high current circuit breaker switchgear contacts, sliding and fixed contacts of isolator switches

Source: Adapted from "Benefits of contact lubricants" by Electrolube

base oils should show a very low tendency to evaporate and a high resistance to oxidation. If temperatures become excessively high for a short time, the lubricant should evaporate or burn away without residues such that no foreign matter (coking) remains, which would interfere with the function of the contact later.

Klüber lubricants are designed and tested for applications in electrical switches and contacts [17,18]. Important issue to resolve in applying a lubricant to electrical switches and contacts is ensuring its compatibility with the contact materials and any surrounding materials. The most popular Klüber lubricants are listed in Table 4.9.

TABLE 4.9 Klüber lubricants for electrical industry

LUBRICANT (COLOR)	COMPOSITION: BASE OIL, THICKENER	NLGI GRADE	WORKING T RANGE °C (°F)	APPLICATION
Amblygon TA 15/2 (beige)	Mineral oil, polyurea	0.5	−20 to 150 (−4 to 302)	Disconnecting contact surfaces
Isoflex Topas NB52 (beige)	Ether + Polyol ester, Ba soap	2	−50 to 120 (−58 to 248)	Rolling and plain bearings, exposed to wide T range, oxidation and corrosive chemicals
Isoflex Topas NB 152 (light beige)	PAO, Ba soap	2	−40 to 140 (−40 to 284)	Rolling and plain bearings, exposed to high speeds and/or T, oxidation and corrosive chemicals
Isoflex Topaz L 152 (light beige)	PAO, Li soap	2	−51 to 149 (−60 to 300)	Roller bearings subject to moderate to high speeds and high temperatures. Highly loaded, moderate- to high-speed bearings
Kluberlectric KR-44-102 (beige), UV indicator	PAO, Li soap	N/A	−40 to 150 (−40 to 302)	Electrical switches, contact (copper, tin, and silver) surfaces

4.2 LUBRICANTS FOR WIND TURBINES

Lubricant manufacturers worldwide develop multiple products based on general recommendations and challenges for lubrication of specific components in wind turbines. With different parts requiring lubrication, multiple products could be required in order to maintain a single turbine. The properties of lubrication products are based on requirements for the greases and pastes to provide long life of the lubricants in various climates and environments where the wind turbines are installed.

The growth of offshore and floating turbines adds to the challenges associated with wind turbine lubrication. Moisture in a gearbox is a big problem because it leads to severe corrosion if not expelled from the system. Therefore, lubricant suppliers need to focus on developing more robust lubricants to help minimize gearbox failure. Wind farm operators also commonly report issues with dirt and water contamination of grease. However, these problems are more common with manual lubrication, versus automatic greasing systems. Proper training of maintenance professionals could help minimize these issues.

Wind turbines require three types of special lubricant: for the hydraulic circuits and brakes, greases for the slewing rings and for bearings. Among the most critical components of the nacelle are the blade bearings, which have to operate under extreme stresses and operating conditions. Improved lubricants and adapted maintenance intervals together with automatic lubricant dispensing systems have contributed to the growth of the latest wind energy technology.

Specialty lithium or calcium soap greases with mineral-, PAO-, or ester-base oils are in use for wind turbines. Synthetic greases should be compatible with the plastic materials involved. In addition, solid lubricants and anti-corrosion additives should provide high wear reduction and less corrosion to make service intervals longer.

4.2.1 Dow Corning Corporation (USA)

Among many lubrication products Dow Corning Corporation manufactures pastes for threaded connections: Molykote®1000 and Molykote® G-Rapid Plus pastes. Typical properties of these pastes are presented in Section 4.1.1, Table 4.1.

With a trend in wind farms moving offshore so as to capture maximized wind speeds and therefore run more efficiently, maintenance will increasingly become more difficult. This emphasizes the importance of proper lubrication at the design stage. To date, very poor maintenance of wind turbine lubrication has been seen in the field. But this could be improved by using the most suitable lubricant for every particular environment and operating condition. For example, if the life of the bearings which operate under high load in a wet, corrosive environment in wind turbine is threatened to shorten, the equipment life can be extended by applying Molykote G-0102 heavy-duty bearing grease, which is well suited in the presence of water.

4.2.2 ExxonMobil (USA)

ExxonMobil produces multiple products for lubrication on different parts of wind turbines. Compared to conventional gear oils, synthetic Mobil gear oils also deliver better wear protection at high temperatures, better oil flow and reduced risk of oil starvation at low temperatures, and reduced traction for improved energy efficiency. Synthetic wind turbine gear oils typically deliver oil drain intervals of three to four years, longer in some cases.

One of the most recommended by Mobil is synthetic gear oil Mobilgear SHC XMP series for use in the main gearbox. This oil provides all-around protection against gear and bearing wear (such as micropitting and scuffing wear). It provides excellent foam and air release, and outstanding stability in the presence of water contamination and corrosion protection. Testing showed 3x improvement in oxidation life with Mobilgear SHC XMP synthetic oil [19].

Another available oil is synthetic hydraulic oil Mobil SHC, which provides maximum anti-wear protection while being shear stable, works in a wide temperature range, and is resistant to deposit formation. Mobil SHC 600 Series synthetic gear and bearing oils are recommended for use in ancillary wind turbine gearboxes. These oils demonstrate excellent high- and low-temperature capability, provide superior wear protection, are highly resistant to oxidation and slugging, and have low traction coefficient.

Mobil also produces a number of synthetic greases for use in open gears, for lubrication of main, pitch, and yaw bearings (Mobil SHC Grease 460WT), and for generator bearings (Mobil SHC 100). For open gears Mobil recommends non-synthetic lubricant Mobiltac 375 NC, which is a nonleaded, diluent-type lubricant for heavily loaded open gears. It provides excellent protection of gear teeth and other machine elements under boundary lubrication conditions. Another product for the same application is Mobilgear OGL 007, which is manufactured based on advanced technology, contains extreme pressure additives and graphite for heavy loads, is lead and chlorine free, and is adhesive in nature; use of this grease reduces environmental impact.

Mobil lubrication products, applications, and major features are listed in Table 4.10 [20,21].

TABLE 4.10 ExxonMobil lubricants for wind turbines

LUBRICATION POINTS	MOBIL PRODUCTS	FEATURES
Main, pitch, yaw and generator bearings	Mobilith SHC™ 100 Mobil SHC™ Grease 460 WT Mobil SHC™ Grease 102 WT	Exceptional high- and low-temperature performance, extreme pressure and anti-wear properties, resistant to water contamination
Gear box	Mobilgear SHC™ XMP 320 Mobil SHC™ Gear 320 WT	Wear, corrosion, antifoam protection, stable during water contamination, long gear, bearing, and oil life
Pitch and yaw gear	Mobilgear SHC™ XMP Series Mobil SHC™ 600 Series	Wear, corrosion, antifoam protection, stable during water contamination, long gear, bearing, and oil life, high- and low-temperature capabilities, resists oxidation and sludging.
Hydraulic System	Mobil SHC™ 524, Mobil DTE 10 Excel™ Series	Wide temperature range, anti-wear protection, highly shear stable, long oil, filter, and component life
Open gear (diluent)	Mobiltac™ 375 NC Non Synthetic	Wear protection, high viscosity, protects gear teeth and machine elements
Open gear (non-diluent)	Mobilgear™ OGL 007	Wear protection, EP additives, and graphite for heavy loads, lead and chlorine free
Ancillary gear boxes	Mobil SHC™ 600 Series	High- and low-temperature capabilities, resists oxidation, sludging, and wear, low traction coefficient

4.2.3 Shell (Netherlands, UK)

Shell manufactures several lubrication products for wind turbines and distributes them through blade bearing European suppliers such as IMO, Liebherr, Rollix, and Rothe Erde. Shell lubricants are used by wind turbine manufacturers including Vestas, Acciona, Gamesa, Dong Fang New Energy Equipment, Sinovel Wind Group, and Siemens Wind Power. Recently Shell introduces wind turbine portfolio to North America [22].

Lubrication products from Shell recommended for use in wind turbines include greases, gearbox oils, and hydraulic fluids.

Shell Rhodina BBZ blade-bearing grease is designed to provide protection to bearings against fretting corrosion, moisture contamination, and false brinelling at temperatures as low as −55°C. This grease is lubricating the blade bearings of many wind turbines globally.

Shell Omala S4 GX 320 synthetic gearbox oil provides excellent protection against common failure modes, including micropitting and bearing wear. Offering excellent low-temperature fluidity and long oil life, Shell Omala S4 GX 320 provides benefits for difficult-to-maintain wind turbine gearboxes.

Shell Gadus S5 V100 2: high-speed low temperature generator bearing grease.

Shell Tellus S4 VX hydraulic oil.

Shell Tellus Arctic 32 is used as the hydraulic fluid for extreme-climate wind turbines, used by wind turbine OEMs including GE Wind, Voith Wind, Vestas, Dongfang Wind Turbines, Sinovel, RePower, Nordex, and DHI. The product has demonstrated its performance in the harsh winters of Mongolia, Scandinavia, and the Americas at temperatures as low as −40°C (−40°F).

In addition, Shell Lubricants also offers Shell Tivela S 150 & 320 synthetic gear oil for yaw and pitch drives; Shell Albida EMS 2 electric motor bearing synthetic grease; Shell Stamina HDS main bearing grease; and Shell Malleus GL & OGH premium quality open gear grease.

4.2.4 Fuchs Petrolub AG (Germany)

Two sister companies Fuchs Europe Schmier-Stoffe and Fuchs Lubritech (subsidiaries of independent lubricant manufacturer Fuchs Petrolub AG) formed Fuchs Windpower Division to manufacture lubricants in the wind power industrial sector. Fuchs Windpower Division has two subdivisions: one (Fuchs Lubritech) manufactures greases and grease pastes and another subdivision (Fuchs) specializes in producing gear and hydraulic oils [23,24].

Oils manufactured by Fuchs are industrial gear lubricants based on poly-alphaolefins (PAO), developed in conjunction with Flender and approved by several leading OEMs such as FAG and Hansen. These lubricants are specially developed for use in environments where extreme temperatures occur, even with short-term peak temperatures up to 150°C (302°F).

Gear oils manufactured by Fuchs:

- *Renolin CLP VCI* is an EP gear oil with excellent corrosion protection properties; it contains specially developed volatile corrosion inhibitors (VCI) components for safe storage and transport of machines and components. Machine elements and gears are also protected against corrosion without direct contact of the oil with the metal surface.
- *Renolin MR 90* has excellent cleaning and flushing properties and very good wear protection; it is a special rig oil providing corrosion protection.
- *Renolin HighGear Synthetic Gear Oil* is an industrial gear oil based on polyalphaolefin and newest additive technology and plastic deformation (PD) technology; this oil is recommended for pre-damaged gear sets and critical applications.
- *Renolin Unisyn CPL* are fully-synthetic industrial gear oils based on PAO with excellent wear protection, good ageing stability, very high viscosity index (VI), and a very low pour point. They provide excellent air release properties, very good demulsibility, and reliable protection against micropitting and white etching cracks. Well-known gear manufacturers approved Renolyn Unisyn CPL 220/320 for use in the wind industry.

Greases manufactured by Fuchs developed for use in bearings:

- *Gleitmo 585 K* is a special lubricant for pitch and yaw bearings (tooth system and bearing). It is a fully-synthetic Li soap heavy-duty grease. This grease contains a synergistic combination of white solid lubricants. This combination offers excellent protection against wear, especially under critical operation conditions like vibration and oscillation movements under high load, which are typical for pitch and yaw bearings on wind turbines. Gleitmo 585 K is well known in the wind power industry and has been used as OEM first fill and service lubricant for many years with best results. It is also recommended for use for the tooth system of the yaw and pitch bearings.

- *Stabyl Eos E 2* is another high-performance grease for wind-power applications (tooth system and bearing). This grease is based on a fully-synthetic ester and a lithium soap as thickener. It fulfills the highest technical requirements for modern lubricants used in wind turbines. Stabyl Eos E 2 was developed in a perennial research project in cooperation with leading bearing manufacturers and is successfully in use on wind turbines as general purpose lubricant. It is also recommended for the tooth system of the yaw- and pitch bearings. Both greases work in wide temperature range −45°C up to 130°C (−49°F to 266°F), suitable for all types of climate conditions; the consistency is NLGI grade 2.

4.2.5 Klüber (Germany)

Companies like Midland, Texas-based Global Wind Power Services, which provides service to the wind power industry, are recommending Klüber lubricants to their customers. Global uses as many as ten different Klüber Lubrication products in the wind turbines it services. Klüber synthetic oils perform well at reducing heat in gear boxes and prolonging the use of the oils.

According to customers [25], these oils keep gears running smoothly and extend the interval between oil changes. Each of the products complies with or exceeds performance parameters stipulated in the standards currently in place.

The Klüber line of synthetic oils includes polyalphaolefin (PAO) oil Klübersynth GEM 4 N; polyglycol (POG) oil Klübersynth GH 6 and rapidly biodegradable ester oil Klübersynth GEM 2.

4.2.6 Total S.A (France)

Total Lubricants manufactures many gear oils, hydraulic and transformers fluids, greases, and coolants specifically designed for wind turbine applications. Total gearbox oils and greases are presented in Tables 4.11 and 4.12 [26].

4.2.7 Castrol (UK)

Castrol offers a full range gearbox oils and greases for lubricating bearings and gear in multiple locations in wind turbines. Castrol oils with anti-wear

TABLE 4.11 Total lubricants: gearbox oils

GEARBOX OILS	TYPE	SPECIAL PROPERTIES
Carter XER 320[a]	Mineral	Protection against teeth micropitting, corrosion protection in sea water
Carter SH 320[b]	PAO	Excellent for very low temperatures, miscible with mineral oil, protection against teeth micropitting
Carter SY WM 320	PAG	Oxidation stability, protection against teeth micropitting
Carter BIO 320	Synthetic Esters	Biogradable > 75% after 28 days, corrosion protection in sea water

Note: Maximum interval between oil changes:
[a] 3 years,
[b] 5 years

additives are listed in Table 4.13 and the greases—in Table 4.14 [27]. Some Castrol greases include Microflux Trans (MFT) additive package, which provides optimum wear protection and an extremely low coefficient of friction even under extremes of pressure, vibration, shock loads, at high or low speeds, or varying operational conditions.

TABLE 4.12 Total lubricants: bearing greases

GREASES	BASE OIL, THICKENER	WORKING T RANGE, °C (°F)	APPLICATION
Multis Complex SHD 100	PAO, Li complex	−50 to 160 (−58 to 320)	Generator bearings, high rotation speeds
Multis Complex SHD 220	PAO, Li complex	−50 to 160 (−58 to 320)	Generator bearings, moderate rotation speeds
Multis Complex SHD 460	PAO Li complex	−40 to 160 (−40 to 320)	Main shaft, pitch and yaw gear, slow-to-moderate rotation speeds
COPAL OGL 0	Mineral oil, Al complex	−20 to 150 (−4 to 302)	Solid lubricant: reduces friction, protects parts from wear

TABLE 4.13 Castrol oils for wind turbines

GEARBOX OILS	BASE OIL	WORKING T RANGE, °C (°F)	APPLICATION
Optigear™ Synthetic X	PAO	−30 to 95 (−22 to 203)	Cylindrical, bevel gears and planetary gears, oil-lubricated rolling bearings
Optigear™ Synthetic A	PAO	−30 to 95 (−22 to 203)	Sliding and rolling bearings, industrial gears at high temperatures and under high mechanical loads
Tribol™ Bio Top 1418	Ester	−25 to 90 (−13 to 194)	Gears, rolling and sliding bearings, as well as in revolving equipment, biodegradable
Tribol™ 1710	Mineral/PAO	−30 to 95 (−22 to 203)	Gear oil with special high-performance additives for wind-power gears
Tribol™ 1100	Mineral	−20 to 90 (−4 to 194)	Modern cylindrical and worm gear pairs, rolling and sliding bearings, revolving equipment
Optigear™ BM	Mineral	−10 to 90 (14 to 194)	Spur and bevel gear units, even under severe operating conditions, worm gear units, rolling and sliding bearings, gear couplings, circulating systems

TABLE 4.14 Castrol greases for wind turbines

GREASES	BASE OIL, THICKENER; NLGI GRADE	WORKING T RANGE, °C (°F)	APPLICATION
Ortitemp™ TT1	Synthetic, Organic/ Inorganic; 1	–60 to 120 (–76 to 248)	Long-term lubrication at high speeds, low T or where large T differences exist
Lonhtime® PD 2	Mineral, Lithium-12-Hydroxystearate; 2	–35 to 140 (–31 to 284)	High-speed rolling and sliding bearings subjected to high mechanical load
Tribol™ 4020	Mineral, Li Complex; 1 and 2	–30 to 150 (–22 to 302)	Sliding and rolling bearings, medium and high mechanical load conditions, water resistant
Tribol™ 3020/1000	Mineral, Li; 000, 00, 0, 1, 2	–40 to 120 (–40 to 248)	Rolling and sliding bearings at low speeds, gears that are not oil-tight, general grease lubrication
Tribol™ GR 1350-2.5 PD (previously called OPTIPIT)	Mineral, Lithium-12-Hydroxystearate, MFT; 2-3	–10 to 140 (14 to 284)	Rolling and sliding bearings, yaw gear, large units with low peripheral speeds in damp, dusty environments, as well as for open gears
Molub-Alloy™ 936 SF HEAVY	Mineral, Mixed Soap, MoS2; 0	–15 to 100 (5 to 212)	Parts with high mechanical loads and at low speeds, water resistant, solvent-free
Molub-Alloy™ 3036/680-1	Mineral, Li; 1	–20 to 120 (–4 to 248)	Rolling and sliding bearings at low speeds and high loads or shock loads in unfavorable ambient conditions

REFERENCES

1. Lutz G. A., Jungk M., Lauer B. 2012. Full Life Wind Turbine Gearbox Lubricating Fluids, Final Report, Submitted by Dow Corning Corporation to United States Dept. of Energy (DOE): 111, February. https://www.osti.gov/servlets/purl/1041556/

2. Wagman D. 2008. Lubrication: Industry Leaders Look to Fix the Varnish Issue, *Power Engineering*, February. http://www.power-eng.com/articles/print/volume-112/issue-2/features/lubrication-industry-leaders-look-to-fix-the-varnish-issue.html

3. Molykote® Industrial Lubricants, Molykote from Dow Corning. http://pdf.directindustry.com/pdf/dow-corning/molykote-industrial-lubricants/21122-2049.html

4. Dow Corning Power & Utility Solutions: There's a lot on the line, 2011. Form No. 80-3459A-01, Dow Corning Corporation, AMPM140-11: 2. https://krayden.com/pdf/molykote_power_utilities_grease_solutions.pdf

5. Specialty Lubricants for the Wind Energy Industry. https://cdn.chempoint.com/ls/LubricantSpecialty/media/LubricantSpecialty/80-3452-01-Pastes-for-Wind-Turbine-Assembly.pdf?ext=.pdf.

6. Design Engineer's Guide to Selecting a Lubricant, 2016. Electric Power Transmission & Distribution, Nye Lubricants, LubeNotes. https://www.nyelubricants.com/site/get_document.php?id=3645

7. Lubricants for Stationary Separable Electrical Connectors, 2012. Nye Lubricants LubeNotes, No. 9. https://www.nyelubricants.com/stuff/contentmgr/files/0/24d7a2dc2442a71906571ee688ad6bf3/en/lubenote_stationary_separable_electrical_connectors.pdf

8. How To Choose The Best Lubricant for Electric Switch Contacts, 1998. Nye Lubricants, LubeLetter, Volume 26, Number 1.

9. Lubricants for Sliding Contacts in Electric Switches, 2016. Nye Lubricants, LubeNotes. https://www.nyelubricants.com/site/get_document.php?id=3722

10. Nye's Family of Anderol® Replacement Lubricants, 2013. Nye Lubricants, LubeNotes. https://www.nyelubricants.com/stuff/contentmgr/files/0/2ab16cba30c7287dd39ff65a9fbf9a04/en/lubenote_anderol.pdf

11. Manufacturer & Distributor of Two Silver Based Products. Cool-Amp. http://www.coolamp.com/index.html

12. Silver filled conductive grease, Chemtronics. http://www.chemtronics.com/products/americas/TDS/Cw7100tds.pdf

13. Conductive Grease and Electrical Contact Lubricant "NO-OX-ID A-Special Electrical Grade". http://www.sanchem.com/aSpecialE.html

14. Electrical Jointing, Tyco Electronics. www.gvk.com.au/pdf/electrical_jointing.pdf

15. High Voltage Contact grease, Contralube, UK. http://highvoltagecontactgrease.com/

16. Benefits of contact lubricants, Electrolube. http://www.contactlubricants.com/the-benefits-of-contact-lubricants.html

17. Keeping contact: specialty lubricants for electrical contacts. 2012. Kluber Lubrication. Bulletin B053001002, Edition 11-2012: 1-11, München, Germany. http://pdf.directindustry.com/pdf/kluber-lubrication/keeping-contact-lubricant/12008-367115.html

18. Klüber lubricants for electrical switches and contacts, 2008. Bulletin B053001002, May Edition, Klüber Lubrication München KG. http://www.industrialbearings. com.au/uploads/catalogs/klueber_elektrische_kontakte-en_1351825422.pdf

19. Daschner E. 2011. Wind Turbine Lubrication Challenges and Increased Reliability, MDA Forum, HANNOVER MESSE. http://files.messe.de/ abstracts/43459_7_Daschner_ExxonMobile(1).pdf

20. Wind Turbine Challenges and considerations. Mobil Industrial Lubricants. http://www.ontario-sea.org/Storage/27/1876_Imperial_Oil_WWEC_Poster_ Paper_Submission.pdf

21. Powering Wind. Powering Progress. 2015. ExxonMobil Industrial Lubricants.

22. Shell Introduces Wind Turbine Lubricant Portfolio to North America, 2011. http://www.machinerylubrication.com/Read/25079/Shell-wind-turbine-lubricant-portfolio

23. Special lubricants for wind power plants, Fuchs Windpower Division. http:// www.fuchs-lubritech.com/fileadmin/downloads/brochures_industries/ br_windpower_e.pdf or http://dneproil.com/pdf_download/Lubricants-for-Windpower-plants.pdf

24. Brazen D. 2009. Lubrication of wind turbines, Presentation at Kansas Renewable Energy Conference, October: 17. www.kcc.ks.gov/energy/kwrec_09/presenta-tions/A2_Brazen.pdf

25. Siebert H. 2006. Wind Turbines Power Up with Oil, "Lubrication & Fluid Power, September-October Issue. http://www.klubersolutions.com/pdfs/_misc/wind_ turbines_power_up_with_oil.pdf

26. Wind energy powered by Total Lubricants. http://www4.total.fr/Europe/ Hungary/PDF/E37_WINDENERGY.pdf

27. Tschauder K., Leather J. 2011. Fundamentals of Lubrication Gear Oil Formulation, NREL Tribology Seminar, Broomfield, CO. http://www.nrel.gov/ wind/pdfs/day1_sessionii_1_castrol_leather.pdf

Addendum 1:
Grease Composition
and Properties

TABLE A1-1 Base oil groups by the American Petroleum Institute (API)

GROUP	SATURATES, %	SULFUR CONTENT, %	VISCOSITY INDEX (SAE)	OIL	PROPERTIES	RELATIVE COST
I	<90	>0.03	80 to 120	Mineral, Solvent refined	Cheapest oils	1.0
II	>90	<0.03	80 to 120	Mineral, hydrotreated	Better anti-oxidation properties and a clearer color than Group I oils	1.05
III	>90	<0.03	Above 120	Mineral, severe hydrocracked	More refined, purer mineral oil	1.5
IV		Polyalphaolefins (PAO), synthetic		Chemical process	Excellent performance over a widerange of lubricating properties	2.5–3
V		All other synthetic oils*				5 to 10+

Note: * Synthetic base oils in Group V include silicone, esters, polyalkylene glycols (PAG), polyolester, alkylated naphthalene, polybutenes, biolubes and others

TABLE A1-2 Grease composition: fluids/oils (from 75% to 95% of volume)

FLUID TYPE	GROUP	GREASE TMAX, °C (°F)	FLUID TYPE	GROUP	GREASE TMAX, °C (°F)
Paraffinic oil	Mineral	149 (300)	Phosphate Ester	SHC	116 (240)
Naphtenic oil	Mineral	149 (300)	Polyalkyne Glycols (PAG)	SHC	149 (300)
Phosphate esters	SHC	116 (240)	Polyglycols	SHC	170 (340)
Diesters	SHC	182 (360)	Polyalphaolefin (PAO)	SHC	132 (270)
Organic esters	SHC	116 (240)	Perfluoropolyester (PFPE)	Fluorocarbons	230 (440)
Polyphenyl ether	SHC	260 (500)	Chlorinated phenyl methyl silicones	Silicones	232 (450)
Polyol Ester	SHC	182 (360)	Methyl silicones	Silicones	232 (450)
Silicate ether or disiloxane	SHC	204 (400)	Phenyl methyl silicones	Silicones	232 (450)

Note: SHC- Synthetic Hydrocarbons; Tmax – maximum operating temperature

TABLE A1-3   Grease composition: thickeners (from 5% to 20% of volume)

THICKENER	GROUP	GREASE TMAX, °C (°F)	THICKENER	GROUP	GREASE TMAX, °C (°F)
Calcium Soap	Soap	93 (195)	Aluminum Complex	Soap	177 (350)
Calcium Complex/ Sulfonate	Soap	177 (350)	Microgel	Nonsoap, inorganic	90 (195)
Calcium 12-Hydroxystearate	Soap	110 (230)	Bentonite (Clay)	Nonsoap, inorganic	177–377 (350–700)
Sodium Soap	Soap	130 (266)	Urea compounds (polyurea)	Nonsoap, organic	177 (350)
Barium Soap	Soap	120 (250)	Polytetrafluoroethylene (PTFE, Teflon)	Nonsoap, organic	360 (680)
Barium Complex Soap	Soap	141 (285)	Fluorine compounds	Nonsoap, organic	N/A
Lithium Soap	Soap	135 (275)	Terephthalate, organic dyes	Nonsoap, organic	N/A
Lithium Complex	Soap	149–177 (300–350)	Carbon black	Nonsoap, inorganic	N/A
Lithium 12-Hydroxystearate	Soap	121 (250)	Silica-gel	Nonsoap, inorganic	N/A
Aluminum Soap	Soap	79	Fumed silica	Nonsoap, inorganic	N/A

TABLE A1-4   Grease composition: additives (from 0% to 15% of volume)

Oxidation inhibitors	Emulsifiers, Demulsifiers	Oxygen scavenging additive
Rust/corrosion inhibitors	Viscosity Index (VI) improvers	Pour point depressants
Anti-wear (AW) agents	Detergents	Tackiness/adhesive agents
Extreme pressure (EP) additives	Dispersants	Solid additives: MoS2, graphite
Acid neutralizers	Viscosity modifiers	Metal deactivator
Anti-foam additives	Oiliness enhancers	Perfumes, dyes

TABLE A1-5   NLGI grease classification

NLGI NUMBER	ASTM WORKED PENETRATION*	CONSISTENCY
000	445–475	Semifluid
00	300–430	Semifluid
0	355–385	Very soft
1	310–340	Soft
2	265–295	Common grease
3	220–250	Semihard
4	175–205	Hard
5	130–160	Very hard
6	85–115	Solid

Source: *Lubrication of Power Plant Equipment*, 1991. Facilities Instructions, Standards, and Techniques (FIST), Vol. 2–4, United States Department of the Interior Bureau of Reclamation (USBR), Springfield, VA: 33. http://www.usbr. gov/power/data/fist/fist2_4/vol2–4.pdf

Note: *   Grease Consistency is measured according to ASTM D217-17 "Standard Test Method for Cone Penetration of Lubricating Grease", depth of cone penetration is measured in 0.1 mm at 25°C

TABLE A1-6   Low temperature (LT) limits for greases

GREASE TYPE	BASE OIL	LT LIMIT, °C (°F)
Conventional industrial greases	Mineral oil	−34 (−30)
Low-temperature greases	Mineral oil	−45.6 (−50)
Synthetic greases	PAO	−54 (−65)
Synthetic greases	Diester	−62 (−80)
Synthetic greases	PFPE	−70 (−94)
Special LT greases	Silicones	−79 (−110)

Addendum 2: Grease Compatibility

TABLE A2-1 Thickeners compatibility chart

	ALUMINUM COMPLEX	BARIUM COMPLEX	CALCIUM STEARATE	CALCIUM 12-HYDROXY	CALCIUM COMPLEX	CALCIUM SULFONATE COMPLEX	LITHIUM STEARATE	LITHIUM 12-HYDROXY	LITHIUM COMPLEX	SODIUM	CLAY (NON-SOAP)	POLYUREA	POLYUREA SHEAR (STABLE)
Aluminum Complex		I	I	C	I	B	I	I	B	I	I	I	C
Barium Complex	I		I	C	I	C	I	I	I	I	I	I	B
Calcium Stearate	I	I		I	I	C	C	B	C	I	C	I	C
Calcium 12-Hydroxy	C	C	I		B	B	C	C	C	I	C	I	C
Calcium Complex	I	I	I	B		I	I	I	C	I	I	C	C
Calcium Sulfonate Complex	B	C	C	B	I		B	B	C	I	I	I	C
Lithium Stearate	I	I	C	C	I	B		I	C	C	I	I	C
Lithium 12-Hydroxy	I	I	B	C	I	B	I		C	I	I	I	C
Lithium Complex	B	I	C	C	C	C	C	C		B	I	I	C
Sodium	I	I	I	I	I	I	C	I	B		I	I	C
Clay (Non-Soap)	I	I	C	C	I	I	I	I	I	I		I	B
Polyurea	I	I	I	I	C	I	I	I	I	I	I		C
Polyurea Shear (Stable)	C	B	C	C	C	C	C	C	C	C	B	C	

Note: Compatibility Rating: C = Compatible; I = Incompatible; B = Borderline

TABLE A2-2   Base oil compatibility chart

BASE OILS	MINERAL	ESTER	PAG*	METHYL SILICONE	PHENYL SILICONE	PPE*	PFAE*
Mineral	–	C	I	I	B	I	I
Ester	C	–	C	I	C	C	I
PAG*	I	C	–	I	I	I	I
Methyl Silicone	I	I	I	–	B	I	I
Phenyl Silicone	B	C	I	B	–	C	I
PPE*	I	C	I	I	C	–	I
PFAE*	I	I	I	I	I		–

Source: Adopted from Grease Compatibility Information. https://www.petroliance.com/
sites/default/files/PDF/Grease/Grease%20Compatibility%20Information.pdf
Note: * PAG- Polyalkylene Glycol; PPE- Polyphenyl Esters; PFAE- Perfluorinated Aliphatic Ester

Addendum 3: Storage and Shelf Life of the Lubricants

TABLE A3-1   Effect of grease components on lubricants storage life

GREASE COMPONENT	INCREASE STORAGE LIFE	DECREASE STORAGE LIFE
Base Oil	Highly refined mineral oils, synthetic hydrocarbons, inert synthetic	Lower grade mineral oils, inorganic esters
Additive	Rust and oxidation inhibitors	EP additives
Thickener	No	Yes

Source: Adopted from Troyer D. and Kucera J. Lubricant storage life limits—industry needs a standard, *Machinery Lubrication*, May 2001

TABLE A3-2   Effect of storage conditions on lubricants life

STORAGE CONDITION	INCREASE STORAGE LIFE	DECREASE STORAGE LIFE
Temperature	Low	High
Temperature Variability	Low	High
Humidity	Low	High
Agitation	Low	High
Container Type	Plastic container or liners	Metal drums, especially poorly conditioned ones
Outdoor Storage	No	Yes

TABLE A3-3   Estimated shelf life of lubricants

OILS	SHELF LIFE, YEARS	GREASES	SHELF LIFE, YEARS	EXCEPTIONS	SHELF LIFE, YEARS
Mineral	5	Mineral	3	Rust Preventive	2
Synthetic	5	Synthetic	3	Open Gear Lubricants	2
Coolants, General	5				

Source: Adopted from Lubricant Storage, Stability, and Estimated Shelf Life, Chevron, USA

Addendum 4: Lubrication Glossary

A

Absolute Viscosity: A term used interchangeably with viscosity to distinguish it from either kinematic viscosity or commercial viscosity. Absolute viscosity is the ratio of shear stress to shear rate. It is a fluid's internal resistance to flow. The common unit of absolute viscosity is the poise. Absolute viscosity divided by fluid density equals kinematic viscosity. It is occasionally referred to as dynamic viscosity. Absolute viscosity and kinematic viscosity are expressed in fundamental units. Commercial viscosity such as Saybolt viscosity is expressed in arbitrary units of time, usually seconds.

Additive: A compound that enhances some property of, or imparts some new property to the base fluid. In some hydraulic fluid formulations, the additive volume may constitute as much as 20% of the final composition. The more important types of additives include anti-oxidants, anti-wear additives, corrosion inhibitors, extreme pressure (EP) additive, viscosity index improvers, and foam suppressants. Other common additives are: demulsifier, detergent, dispersant, fluidizer, oiliness agent, pour-point depressant, rust inhibitor, tackiness agent, and viscosity index (VI) improver.

Additive Level: The total percentage of all additives in an oil. Expressed in % of mass [weight] or % of volume.

Additive Stability: The ability of additives in the fluid to resist changes in their performance during storage or use.

Adhesion: The property of a lubricant that causes it to cling or adhere to a solid surface.

ANSI: American National Standards Institute.

Anti-foam Agent: One of two types of additives used to reduce foaming in petroleum products: silicone oil to break up large surface bubbles, and various kinds of polymers that decrease the number of small bubbles entrained in the oils.

Anti-oxidants: Prolong the induction period of a base oil in the presence of oxidizing conditions and catalyst metals at elevated temperatures. The additive is consumed and degradation products increase not only with increasing and sustained temperature, but also with increases in mechanical agitation or turbulence and contamination (water, metallic particles, and dust).

Anti-seize Compound: Grease-like substance containing graphite, moly, or metallic solids (copper, zinc, silver, or lead), which is applied to threaded joints, particularly those subjected to high temperatures, to facilitate easy separation when required.

Anti-static Additive: An additive that increases the conductivity of a hydrocarbon fuel to hasten the dissipation of electrostatic charges during high-speed dispensing, thereby reducing the fire/explosion hazard.

Anti-wear Additives: Improve the service life of tribological elements operating in the boundary lubrication regime. Anti-wear compounds (for example, ZDDP and TCP) start decomposing at 90°C to 100°C and even at a lower temperature if water (25 to 50 ppm) is present.

Apparent Viscosity: Introduced to distinguish between the viscosity of oil and grease; the viscosity of grease is referred to as "apparent viscosity." Apparent viscosity is the viscosity of a grease that holds only for the shear rate and temperature at which the viscosity is determined. At start-up, grease has resistance to motion, implying high viscosity. However, as grease is sheared between wearing surfaces and moves faster, its resistance to flow reduces. Its viscosity decreases as the rate of shear increases. By contrast, oil at a constant temperature would have the same viscosity at start-up as what it has when moving.

ASTM: American Society for Testing Materials, a society for developing standards for materials and test methods.

B

Base Oil: A base oil is a base stock or blend of base stocks used in an API-licensed engine oil.

Base Stock: The base fluid, usually a refined petroleum fraction or a selected synthetic material, into which additives are blended to produce finished lubricants.

Biodegradable: Capable of being broken down chemically, especially into innocuous products, by the action of living microorganisms in the environment.

Bleeding: The separation of some of the liquid phase from a grease.

Blending: The process of mixing lubricants or components for the purpose of obtaining the desired physical and/or chemical properties.

Boundary Lubrication: Form of lubrication between two rubbing surfaces without development of a full-fluid lubricating film. Boundary lubrication can be made more effective by including additives in the lubricating oil that provide a stronger oil film, thus preventing excessive friction and possible scoring.

C

Carbon Residue: Coked material formed after lubricating oil has been exposed to high temperatures. Many consider the type of carbon formed to be of greater significance that the quantity.

Centipoise (cp): A unit of absolute viscosity. 1 centipoise = 0.01 poise.

Centistoke (cst): A unit of kinematic viscosity. 1 centistoke = 0.01 stoke.

Chemical Stability: The tendency of a substance or mixture to resist chemical change.

Cleanable Filter: A filter element which, when loaded, can be restored by a suitable process, to an acceptable percentage of its original dirt capacity.

Cleanliness Level: A measure of relative freedom from contaminants.

Complex Grease: A lubricating grease thickened by a complex soap consisting of a normal soap and a complexing agent.

Consistency: The degree to which a semisolid material such as grease resists deformation. Sometimes used qualitatively to denote viscosity of liquids.

Contaminant: Any foreign or unwanted substance that can have a negative effect on system operation, life or reliability.

Corrosion: The decay and loss of a metal due to a chemical reaction and its environment. It is a transformation process in which the metal passes from its elemental form to a combined (or compound) form.

Corrosion Inhibitor: Additive for protecting lubricated metal surfaces against chemical attack by water or other contaminants. There are several types of corrosion inhibitors. Polar compounds wet the metal surface preferentially, protecting it with a film of oil. Other compounds may absorb water by incorporating it in a water-in-oil emulsion so that only the oil touches the metal surface. Another type of corrosion inhibitor combines chemically with the metal to present a nonreactive surface.

D

Degradation: The progressive failure of a machine or lubricant.

Demulsifier: An additive that promotes oil–water separation in lubricants that are exposed to water or steam.

Deplete: The depletion of additives expressed as an approximate percentage.

Deposits: Oil-insoluble materials that result from oxidation and decomposition of lube oil and contamination from external sources and engine blow-by. These can settle out on machine or engine parts. Examples are sludge, varnish, lacquer, and carbon.

Detergent: In lubrication, a detergent is either an additive or a compounded lubricant having the property of keeping insoluble matter in suspension, thus preventing its deposition where it would be harmful. A detergent may also redisperse deposits already formed. Detergent additives contain metallic derivatives, such as barium, calcium, and magnesium sulfonates, phosphonate, thiophosphonate, phenate, or salicylate. Because of its metallic composition, a detergent leaves a slight ash when the oil is burned. A detergent is normally used in conjunction with a dispersant.

Diester: A category of base stocks typically used in reciprocating air compressors and vane compressors. Diesters mix with most glycols, polyalphaolefins (PAOs), and other fluids. Diesters offer excellent resistance to oxidation, varnish, and carbon for a long fluid life. Their water separability is inferior to other synthetics, and their low viscosity indexes diminish cold-weather performance.

Dispersant: In lubrication, a term usually used interchangeably with detergent. An additive, usually nonmetallic ("ashless"), which keeps fine particles of insoluble materials in a homogeneous solution. Hence, particles are not permitted to settle out and accumulate. Dispersants are normally used in conjunction with detergents.

Dropping Point: The temperature at which grease becomes fluid enough to drip. The dropping point is an indicator of the heat resistance of grease. As grease temperature rises, penetration increases until the grease liquefies and the desired consistency is lost. The dropping point indicates the upper temperature limit at which grease retains its structure, not the maximum temperature at which grease may be used. A few greases have the ability to regain their original structure after cooling down from the dropping point.

Dry Lubricant: Solid material left between two moving surfaces to prevent metal-to-metal contact, thus reducing friction and wear. Such

materials are especially useful in the region of boundary lubrication, and for lubrication under special conditions of extremely high
or low temperature where usual lubricants are inadequate. They may
be applied in the form of a paste or solid stick, or by spraying, dipping, or brushing in an air-drying carrier which evaporates leaving a
dry film. Some examples of dry lubricants are: graphite, molybdenum
disulfide (moly), boron nitride, and plastics such as tetrafluorethylene
resins (PTFE or Teflon).

Dry Lubrication: The situation when moving surfaces have no liquid lubricant between them.

E

Electrical Insulating Oil: A high-quality oxidation-resistant oil refined to
give long service as a dielectric and coolant for electrical equipment,
most commonly transformers. An insulating oil must resist the effects
of elevated temperatures, electrical stress, and contact with air, which
can lead to sludge formation and loss of insulation properties. It must
be kept dry, as water is detrimental to dielectric strength.

Emulsibility: The ability of a nonwater-soluble fluid to form an emulsion with
water.

Emulsifier: An additive that promotes the formation of a stable mixture, or
emulsion, of oil and water. Common emulsifiers are metallic soaps,
certain animal and vegetable oils, and various polar compounds.

Environmental: All material and energy present in and around an operating contaminant system, such as dust, air moisture, chemicals, and
thermal energy.

Environmental Contaminant: All material and energy present in and around
an operating system, such as dust, air moisture, chemicals, and thermal energy.

Ester: A chemical compound typically formed through the reaction between
an acid and an alcohol. Many chemically different "esters" are used
for various reasons as either "additives" or "base stocks" for lubricants due to their usually excellent lubricity.

Extreme Pressure (EP) Additive: Lubricant additive that prevents sliding
metal surfaces from seizing under conditions of extreme pressure.
At the high local temperatures associated with metal-to-metal contact, an EP additive combines chemically with the metal to form a
surface film that prevents the welding of opposing asperities and the

consequent scoring that is destructive to sliding surfaces under high loads. Reactive compounds of sulfur, chlorine, or phosphorus are used to form these inorganic films.

Extreme Pressure (EP) Lubricants: Lubricants that impart to rubbing surfaces the ability to carry appreciably greater loads than would be possible with ordinary lubricants without excessive wear or damage.

F

Film Strength: Property of a lubricant that acts to prevent scuffing or scoring of metal parts.

Filter: Any device or porous substance used as a strainer for cleaning fluids by removing suspended matter.

Filter Efficiency: Method of expressing a filter's ability to trap and retain contaminants of a given size.

Filtration: The physical or mechanical process of separating insoluble particulate matter from a fluid, such as air or liquid, by passing the fluid through a filter medium that will not allow the particulates to pass through it.

Fire point: The temperature at which oil will burn if ignited.

Flash point: The temperature at which oil gives off ignitable vapors. The flash point is not necessarily a safe upper limit for oil because some decomposition takes place below the flash point.

Foam Inhibitor: A substance introduced in a very small proportion to a lubricant or a coolant to prevent the formation of foam due to aeration of the liquid, and to accelerate the dissipation of any foam that may form.

Foaming: A frothy mixture of air and a petroleum product (e.g., lubricant, fuel oil) that can reduce the effectiveness of the product and cause sluggish hydraulic operation, air binding of oil pumps, and overflow of tanks or sumps. Foaming can result from excessive agitation, improper fluid levels, air leaks, cavitation, or contamination with water or other foreign materials. Foaming can be inhibited with an anti-foam agent. The foaming characteristics of a lubricating oil can be determined by blowing air through a sample at a specified temperature and measuring the volume of foam, as described in test method ASTM D 892.

Fretting Corrosion: Can take place when two metals are held in contact and subjected to repeated small sliding, relative motions. Other names for this type of corrosion include wear oxidation, friction oxidation, chafing, and brinelling.

Friction: Resisting force encountered at the common boundary between two bodies when, under the action of an external force, one body moves or tends to move relative to the surface of the other.

G

Gear Oil: A high-quality oil with good oxidation stability, load-carrying capacity, rust protection, and resistance to foaming for service in gear housings and enclosed chain drives. Specially formulated industrial EP gear oils are used where highly loaded gear sets or excessive sliding action (as in worm gears) is encountered.

Graphite: A crystalline form of carbon having a laminar structure, which is widely used as a lubricant, either alone as a dry lubricant or added to conventional lubricants, both oils and greases. It may be of natural or synthetic origin.

Grease: A type of lubricant composed of a fluid (typically lubricant oils) thickened with a material that contributes a degree of plasticity (typically soaps). Just as viscosity is the basic property of lubricating oil, consistency is the basic property of grease. Consistency is measured in terms of penetration, tested in terms of tenths of a millimeter that a standard cone acting under the influence of gravity penetrates the sample under controlled test conditions. The greater penetration, the softer the grease.

H

Hydraulic Fluid or Oil: Fluid serving as the power transmission medium in a hydraulic system. The most commonly used fluids are petroleum oils, synthetic lubricants, oil–water emulsions, and water–glycol mixtures. The principal requirements of a premium hydraulic fluid are proper viscosity, high viscosity index, anti-wear protection (if needed), good oxidation stability, adequate pour point, good demulsibility, rust inhibition, resistance to foaming, and compatibility with seal materials. Anti-wear oils are frequently used in compact, high-pressure, and capacity pumps that require extra lubrication protection.

Hydrocarbons: A compound of hydrogen and carbon of which petroleum products are typically examples. Petroleum oils are generally grouped into two parts: Naphthenics possess a high proportion of unsaturated cyclic molecules; paraffinics possess a low proportion of unsaturated cyclic molecules.

I, K

Immiscible: Incapable of being mixed without separation of phases. Water and petroleum oil are immiscible under most conditions, although they can be made miscible with the addition of an emulsifier.

Incompatible Fluids: Fluids which when mixed in a system, will have a deleterious effect on that system, its components, or its operation.

Industrial Lubricant: Any petroleum or synthetic-base fluid or grease commonly used in lubricating industrial equipment, such as gears, turbines, and compressors.

Ingested Contaminants: Environmental contaminant that ingresses due to the action of the system or machine.

Ingression Level: Particles added per unit of circulating fluid volume.

Inhibitor: Any substance that slows or prevents such chemical reactions as corrosion or oxidation.

Insolubles: Particles of carbon or agglomerates of carbon and other materials. Indicates deposition or dispersant drop-out in an engine. Not serious in a compressor or gearbox unless there has been a rapid increase in these particles.

ISO: International Standards Organization, sets viscosity reference scales.

ISO Viscosity: A number indicating the nominal viscosity of an industrial fluid grade at 40°C/104°F as defined by ASTM Standard Viscosity System for Industrial Fluid Lubricants D 2422 (identical to ISO Standard 3448).

Karl Fischer Reagent Method: The standard laboratory test to measure the water content of mineral base fluids. In this method, water reacts quantitatively with the Karl Fischer reagent. This reagent is a mixture of iodine, sulfur dioxide, pyridine, and methanol. When excess iodine exists, electric current can pass between two platinum electrodes or plates. The water in the sample reacts with the iodine. When the water is no longer free to react with iodine, an excess of iodine depolarizes the electrodes, signaling the end of the test.

Kinematic Viscosity: The time required for a fixed amount of an oil to flow through a capillary tube under the force of gravity. The unit of kinematic viscosity is a stoke or centistoke (1/100 of a stoke). Kinematic viscosity may be defined as the quotient of the absolute viscosity in centipoises divided by the specific gravity of a fluid, both at the same temperature.

L, M, N

Lithium Grease: The most common type of grease today, based on lithium soaps.

Load-carrying: Property of a lubricant to form a film on the lubricated surface, which resists rupture under given load capacity conditions. Expressed as the maximum load the lubricated system can support without failure or excessive wear.

Lubricant: Any substance interposed between two surfaces in relative motion for the purpose of reducing the friction and/or the wear between them.

Lubrication: The control of friction and wear between two moving, touching surfaces by placing a friction-reducing substance between them.

Lubricity: Ability of an oil or grease to lubricate; also called film strength.

Material Safety Data Sheet (MSDS): A publication containing health and safety information on a hazardous product (including petroleum). The OSHA Hazard Communication Standard requires that an MSDS be provided by manufacturers to distributors or purchasers prior to or at the time of product shipment. An MSDS must include the chemical and common names of all ingredients that have been determined to be health hazards if they constitute 1% or greater of the product's composition (0.1% for carcinogens). An MSDS also included precautionary guidelines and emergency procedures.

Metal Oxides: Oxidized ferrous particles which are very old or have been recently produced by conditions of inadequate lubrication. Trend is important.

Mineral Oil: Oil derived from a mineral source, such as petroleum.

Miscible: Capable of being mixed in any concentration without separation of phases; e.g., water and ethyl alcohol are miscible.

Mixed Film: A type of lubrication that features a combination of full-film and thin-film elements.

Molybdenum Disulfide, Moly or Molysulfide: A dark green to black, lustrous powder (MoS_2) that serves as a dry-film lubricant in certain high-temperature and high-vacuum applications. It is also used in the form of pastes to prevent scoring when assembling press-fit parts, and as an additive to impart residual lubrication properties to oils and greases.

Multigrade Oil: Multigrade is a term used to describe an oil for which the viscosity/temperature characteristics are such that its low-temperature and high-temperature viscosities fall within the limits of two different SAE numbers. (SAE Standard J300). Multigrade oil may be suitable for use over a wider temperature range than a single-grade oil.

NLGI: The National Lubricating Grease Institute

NLGI Grade: The NLGI consistency number (sometimes called "NLGI grade") expresses a measure of the relative hardness of a grease used for lubrication, as specified by the standard classification of lubricating grease established by NLGI.

O

Oil: A greasy, unctuous liquid of vegetable, animal, mineral, or synthetic origin.

Oil Analysis: The routine activity of analyzing lubricant properties and suspended contaminants for the purpose of monitoring and reporting timely, meaningful, and accurate information on lubricant and machine condition.

Oil Change: The act of replacing dirty oil with clean oil.

Oil Oxidation: Occurs when oxygen attacks petroleum fluids. The process is accelerated by heat, light, metal catalysts, and the presence of water, acids, or solid contaminants. It leads to increased viscosity and deposit formation.

Oiliness: That property of a lubricant that produces low friction under conditions of boundary lubrication. The lower the friction, the greater the oiliness.

Oiliness Agent: A polar compound used to increase the lubricity of a lubricating oil and aid in preventing wear and scoring under conditions of boundary lubrication.

Oxidation: Occurs when oxygen attacks petroleum fluids. The process is accelerated by heat, light, metal catalysts, and the presence of water, acids, or solid contaminants. It leads to increased viscosity and deposit formation. Oxidation results in formation of sludges, varnishes, and gums that can impair equipment operation. The organic acids formed from oxidation are corrosive to metals. Oxidation resistance of a product can be improved by careful selection of base stocks, special refining methods, and addition of oxidation inhibitors. Also, oxidation can be minimized by good maintenance of oil and equipment to prevent contamination and excessive heat.

Oxidation Inhibitor: Substance added in small quantities to a petroleum product to increase its oxidation resistance, thereby lengthening its service or storage life; also called antioxidant. An oxidation inhibitor may work in one of these ways: (1) by combining with and modifying peroxides (initial oxidation products) to render them harmless, (2) by decomposing the peroxides, or (3) by rendering an oxidation catalyst inert.

Oxidative Stability: Stability of a lubricant to resist a chemical union with oxygen. The reaction of grease with oxygen produces insoluble gum, sludges, and lacquer-like deposits that cause sluggish operation, increased wear, and reduction of clearances. Prolonged high-temperature exposure accelerates oxidation in greases.

P

PAO Synthetic Fluid: Polyalphaolefins (PAO), often called synthetic hydro-carbons, are probably the most common type of synthetic base oil used today. They are moderately priced, provide excellent performance, and have few negative attributes. PAO base oil is similar to mineral oil. The advantage comes from the fact that it is built, rather than extracted and modified, making it more pure. Practically all of the oil molecules are the same shape and size and are completely saturated. The potential benefits of PAOs are improved oxidative and thermal stability, excellent demulsibility and hydrolytic stability, a high VI, and very low pour point. Most of the properties make PAOs a good selection for temperature extremes—both high operating temperatures and low start-up temperatures. Typical applications for PAOs are engine oils, gear oils, and compressor oils. The negative attributes of PAOs are the price and poor solubility. The low inherent solubility of PAOs creates problems for formulators when it comes to dissolving additives. Likewise, PAOs cannot suspend potential varnish-forming degradation by-products, although they are less prone to create such material.

Particle Counting: A microscopic technique that enables the visual counting of particles in a known quantity of fluid. The count identifies the number of particles present greater than a particular micron size per unit volume of fluid often stated as particles >10 microns per milliliter.

Particle Density: An important parameter in establishing an entrained particle's potential to impinge on control surfaces and cause erosion.

Particulates: Particles made up of a wide range of natural materials (e.g., pollen, dust, resins), combined with man-made pollutant (e.g., smoke particles, metallic ash); in sufficient concentrations, particulates can be a respiratory irritant.

Penetration: Consistency of a lubricating grease, expressed as the distance in millimeters that a standard needle or cone penetrates vertically into a sample of the material under known conditions of loading, time, and temperature.

Petrochemical: Any chemical substance derived from crude oil or its products, or from natural gas. Some petrochemical products may be identical to others produced from other raw materials such as coal and producer gas.

Polyalkylene Glycol (PAG): Mixtures of condensation polymers of ethylene oxide and water. They are any of a family of colorless liquids with high molecular weight that are soluble in water and in many organic solvents. They are used in detergents and as emulsifiers and plasticizers. PAG-based lubricants are used in diverse applications where petroleum oil-based products do not provide the desired performance—and because they are fire-resistant and will not harm workers or the environment.

Polyglycols: Polymers of ethylene or propylene oxides used as a synthetic lubricant base. Properties include very good hydrolytic stability, high viscosity index (VI), and low volatility. Used particularly in water emulsion fluids.

Polymerization: The chemical combination of similar-type molecules to form larger molecules.

Polyol Ester (POE): A synthetic lubricant base, formed by reacting fatty acids with a polyol (such as a glycol) derived from petroleum. Properties include good oxidation stability at high temperatures and low volatility. Used in formulating lubricants for turbines, compressors, jet engines, and automotive engines.

Polyolefin: A polymer derived by polymerization of relatively simple olefins. Polyethylene and polyisoprene are important polyolefins.

Pour Point: The lowest temperature at which oil flows and is most critical in low-temperature applications. Formation of wax crystals causes flow failure in paraffinic oils.

Pumpability: The ability of a lubricating grease to flow under pressure through the line, nozzle, and fitting of a grease-dispensing system. More practically, pumpability is the ease with which a pressurized grease can flow through lines, nozzles, and fittings of grease-dispensing systems.

R, S

Refining: A process of reclaiming used lubricant oils and restoring them to a condition similar to that of virgin stocks by filtration, clay adsorption, or more elaborate methods.

Reversibility: Ability of grease to return to its original consistency after it is exposed to high temperature or high shock loading for a short period of time. When a grease encounters abnormally high temperatures for short periods of time and then returns to normal operating temperatures, or encounters high shock loading conditions, bleeding of the base oil from the grease may occur. The grease with high reversibility must have the ability to recapture its bases in order to return to its original consistency.

Rust Inhibitor: A type of corrosion inhibitor used in lubricants to protect surfaces against rusting.

Rust Preventive: Compound for coating metal surfaces with a film that protects against rust; commonly used for the preservation of equipment in storage. The base material of a rust preventive may be a petroleum oil, solvent, wax, or asphalt, to which a rust inhibitor is added. A formulation consisting largely of a solvent and additives is commonly called a *thin-film rust preventive* because of the thin coating that remains after evaporation of the solvent. Rust preventives are formulated for a variety of conditions of exposure, e.g., short-time "in-process" protection, indoor storage, exposed outdoor storage, etc.

SAE: Society of Automotive Engineers, an organization serving the automotive industry.

SAE Viscosity: The viscosity classification of a motor oil according to the system developed by the Society of Automotive Engineers and now in general use. "Winter" grades are defined by viscosity measurements at low temperatures and have "W" as a suffix, while "Summer" grades are defined by viscosity at 100°C and have no suffix. Multigrade oils meet both a winter and a summer definition and have designations such as SAE 10W-30, etc.

Saybolt Universal Viscosity (SUV) or Saybolt Universal Seconds, (SUS): The time in seconds required for 60 cubic centimeters of a fluid to flow through the orifice of the standard Saybolt universal viscometer at a given temperature under specified conditions.

Shear Stability: Grease consistency may change as it is mechanically worked or sheared between wearing surfaces. A grease's ability to maintain its consistency when it is worked is its shear stability or mechanical stability. A grease that softens as it is worked is called thixotropic. Greases that harden when they are worked are called rheopectic.

Slumpability or Feedability: Ability of a lubricant to be drawn into (sucked into) a pump. Fibrous greases tend to have good feedability but poor pumpability. Buttery textured greases tend to have good pumpability but poor feedability.

Stoke (St): Kinematic measurement of a fluid's resistance to flow defined by the ratio of the fluid's dynamic viscosity to its density.

Straight Mineral Oil: Petroleum oil containing no additives. Straight mineral oils include such diverse products as low-cost once-through lubricants and thoroughly refined white oils. Most high-quality lubricants, however, contain additives.

Surfactant: May increase the oil's affinity for metals and other materials.

Synthetic Hydrocarbon: A category of base stocks commonly used in engine oils and other automotive applications, hydraulic fluids, air compressors, pump fluids, and chain and gear fluids. They offer excellent thermal and oxidative stability, lubricity and water separation, but they will not mix with glycols. Many (but not all) synthetic oils offer immense advantages in terms of high-temperature stability and low-temperature fluidity, but are more costly than mineral oils. Major advantage of all synthetic oils their chemical uniformity that does not significantly vary from batch to batch.

Synthetic Lubricant: A lubricant produced by chemical synthesis rather than by extraction or refinement of petroleum to produce a compound with planned and predictable properties.

Synthetic Oils: Oils produced by synthesis (chemical reaction) rather than by extraction or refinement. Many (but not all) synthetic oils offer immense advantages in terms of high-temperature stability and low-temperature fluidity, but are more costly than mineral oils. Major advantage of all synthetic oils is their chemical uniformity.

T

Tackiness Agent: An additive used to increase the adhesive properties of a lubricant, improve retention, and prevent dripping and splattering.

Tacky: A descriptive term applied to lubricating oils and greases which appear particularly sticky or adhesive.

Teflon (PTFE): The original form of Teflon® is polytetrafluoroethylene, or PTFE. The molecular structure of Teflon® is based on a chain of carbon atoms, the same as all polymers. Unlike some other fluoropolymers, in Teflon® this chain is completely surrounded by fluorine atoms. The bond between carbon and fluorine is very strong, and the fluorine atoms shield the vulnerable carbon chain. This unusual structure gives Teflon® its unique properties. In addition to its extreme slipperiness, it is inert to almost every known chemical. Teflon® is a registered trademark of the *DuPont Company* for many of its fluorine-based products including fluoropolymer resins, films, coatings, additives, and other products.

Thermal Stability: The ability to resist chemical degradation under high-temperature operating conditions.

Timken EP Test: Measure of the extreme-pressure properties of a lubricating oil. The test utilizes a Timken machine, which consists of a stationary block pushed upward, by means of a lever arm system, against the rotating outer race of a roller bearing, which is lubricated by the product under test. The test continues under increasing load (pressure) until a measurable wear scar is formed on the block.

Timken OK Load: The heaviest load that a test lubricant will sustain without scoring the test block in the Timken Test procedures, ASTM Methods D 2509 (greases) and D 2782 (oils).

Thin Film Lubrication: A condition of lubrication in which the film thickness of the lubricant is such that the friction between the surfaces is determined by the properties of the surfaces as well as by the viscosity of the lubricant.

Tribology: The science and technology of interacting surfaces in relative motion, including the study of lubrication, friction, and wear. Tribological wear is wear that occurs as a result of relative motion at the surface.

Turbine Oil: A top-quality rust- and oxidation-inhibited (R&O) oil that meets the rigid requirements traditionally imposed on steam-turbine lubrication. Quality turbine oils are also distinguished by good demulsibility, a requisite of effective oil–water separation. Turbine oils are widely used in other exacting applications for which long service life and dependable lubrication are mandatory, such as compressors, hydraulic systems, gear drives, and other equipment. Turbine oils can also be used as heat transfer fluids in open systems, where oxidation stability is of primary importance.

V

Varnish: When applied to lubrication, a thin, insoluble, non-wipeable film deposit occurring on interior parts, resulting from the oxidation and polymerization of fuels and lubricants. Can cause sticking and malfunction of close-clearance moving parts. Similar to, but softer, than lacquer.

Viscosity: Measurement of a fluid's resistance to flow. The common metric unit of absolute viscosity is the poise, which is defined as the force in dynes required to move a surface 1 cm^2 in area past a parallel surface at a speed of 1 cm/s, with the surfaces separated by a fluid film 1 cm thick. In addition to kinematic viscosity, there are other methods for determining viscosity, including the Saybolt universal viscosity (SUV), Saybolt Furol viscosity, Engier viscosity, and Redwood viscosity. Since viscosity varies inversely with temperature, its value is meaningless until the temperature at which it is determined is reported.

Viscosity Grade: Any of a number of systems which characterize lubricants according to viscosity for particular applications, such as industrial oils, gear oils, automotive engine oils, automotive gear oils, and aircraft piston engine oils.

Viscosity Index (VI): A commonly used measure of a fluid's change of viscosity with temperature. The higher the viscosity index, the smaller the relative change in viscosity with temperature.

Viscosity Index Improvers (VII): Additives that increase the viscosity of the fluid throughout its useful temperature range. Such additives are polymers that possess thickening power as a result of their high molecular weight and are necessary for the formulation of multigrade engine oils.

Viscosity Modifier: Lubricant additive, usually a high-molecular-weight polymer that reduces the tendency of an A148 viscosity to change with temperature.

Viscosity-temperature Relationship: The manner in which the viscosity of a given fluid varies inversely with temperature. Because of the mathematical relationship that exists between these two variables, it is possible to predict graphically the viscosity of a petroleum fluid at any temperature within a limited range if the viscosities at two other temperatures are known. The charts used for this purpose are the ASTM Standard Viscosity-Temperature Charts for liquid Petroleum Products, available in six ranges. If two known viscosity-temperature points of a fluid are located on the chart and a straight line drawn through them, other viscosity-temperature values of the fluid will fall on this line; however, values near or below the cloud point of the oil may deviate from the straight-line relationship.

Viscous: Possessing viscosity. Frequently used to imply high viscosity.

W, Z

Wear: The attrition or rubbing away of the surface of a material as a result of mechanical action. Various forms of wear are: (1) *abrasive wear*—removal of materials from surfaces in relative motion by cutting or abrasive action of a hard particle (usually a contaminant); (2) *adhesive wear* (scuffing)—removal of materials from surfaces in relative motion as a result of surface contact; (3) *corrosive wear*—removal of materials by chemical action.

Wear Debris: Particles that are detached from machine surfaces as a result of wear and corrosion. Also known as wear particles.

Wear Inhibitor: An additive that protects the rubbing surfaces against wear, particularly from scuffing, if the hydrodynamic film is ruptured.

ZDDP: Zinc dialkyldithiophosphate, an anti-wear additive found in many types of hydraulic and lubricating fluids.

Index